# Data Processing Handbook for Complex Biological Data Sources

# Data Processing Handbook for Complex Biological Data Sources

Edited by

**Gauri Misra**

Amity Institute of Biotechnology, Amity University, Noida, India

ACADEMIC PRESS

An imprint of Elsevier

Academic Press is an imprint of Elsevier
125 London Wall, London EC2Y 5AS, United Kingdom
525 B Street, Suite 1650, San Diego, CA 92101, United States
50 Hampshire Street, 5th Floor, Cambridge, MA 02139, United States
The Boulevard, Langford Lane, Kidlington, Oxford OX5 1GB, United Kingdom

**British Library Cataloguing-in-Publication Data**
A catalogue record for this book is available from the British Library

**Library of Congress Cataloging-in-Publication Data**
A catalog record for this book is available from the Library of Congress

ISBN: 978-0-12-816548-5

For Information on all Academic Press publications
visit our website at https://www.elsevier.com/books-and-journals

Working together
to grow libraries in
developing countries

www.elsevier.com • www.bookaid.org

*Publisher:* Stacy Masucci
*Acquisition Editor:* Rafael E. Teixeira
*Editorial Project Manager:* Rebeka Henry
*Production Project Manager:* Poulouse Joseph
*Cover Designer:* Mark Rogers

Typeset by MPS Limited, Chennai, India

# Dedication

**Dedicated to**
**The lotus feet of Shri Radha Krishna**
**and**
**Goverdhannathji**

# Contents

# List of contributors

**Usha Agrawal**
CRIB Lab, ICMR-National Institute of Pathology, New Delhi, India

**Mohd Tashfeen Ashraf**
School of Biotechnology, Gautam Buddha University, Noida, India

**Marcio Chaim Bajgelman**
Brazilian Biosciences Laboratory, National Center for Research in Energy and Materials, Campinas, Brazil

**Ankur Baliyan**
NISSAN Analysis and Research Center, Yokosuka, Japan

**Bhaswati Banerjee**
School of Biotechnology, Gautam Buddha University, Noida, India

**Nar Singh Chauhan**
Department of Biochemistry, Maharshi Dayanand University, Rohtak, India

**Sudhir K. Gupta**
Faculty of Chemical Sciences, Department of Chemistry, Harcourt Butler Technical University (Formerly HBTI), Kanpur, India

**Hideto Imai**
NISSAN Analysis and Research Center, Yokosuka, Japan

**Gagandeep Jhingan**
Valerian Chem Private Limited, New Delhi, India

**Mahima Khandelwal**
School of Chemical Engineering, University of Ulsan, South Korea

**Amit Kumar**
ICMR-AIIMS Computational Genomics Centre, Convergence Block, All India Institute of Medical Sciences, New Delhi, India

**Atul Kumar**
Amity Institute of Engineering & Technology, Amity University, Greater Noida, India

**Vinit Kumar**
Amity Institute of Molecular Medicine and Stem Cell Research, Amity University, Noida, India

**Gauri Misra**
Amity Institute of Biotechnology, Amity University, Noida, India

**Dibyabhaba Pradhan**
ICMR-AIIMS Computational Genomics Centre, Convergence Block, All India Institute of Medical Sciences, New Delhi, India

**Jyotika Rajawat**
Molecular & Human Genetics Laboratory, Department of Zoology, University of Lucknow, Lucknow, India

**Reshma Rani**
Amity Institute of Biotechnology, Amity University, Noida, India

**Harpreet Singh**
ICMR-AIIMS Computational Genomics Centre, Convergence Block, All India Institute of Medical Sciences, New Delhi, India

**Vijay Kumar Srivastava**
Amity Institute of Biotechnology, Amity University Rajasthan, Jaipur, India

**Leonardo Vazquez**
Key Laboratory of Pathogenic Microbiology and Immunology, Institute of Microbiology, Chinese Academy of Sciences, Beijing, China

**Rupali Yadav**
Dr. Reddy's Institute of Life Sciences, University of Hyderabad Campus, Hyderabad, India

# Foreword

The recent explosion of data in the biological sciences presents both clear opportunities and considerable challenges. Much of these data arise from high-throughput sequencing and associated computations. However, once a protein sequence is determined, its three-dimensional structure and interactions must be characterized, in a process that experimentally is much less high-throughput. The present book melds these two approaches in an extensive description of structural and genomic techniques. This is the third book by Editor Gauri Misra published by an international publisher. Her earlier works have been well accepted in scientific community and we hope she continues her good work enriching technical scientific literature resources.

Some of the work in the book treats topics related to the generation of data. A clear example of this is the chapter by Pradhan et al. on methods for high-throughput sequencing. Similarly high-throughput data streams can also result from the collection of mass spectroscopy data in the identification of biomolecules and protein:protein interactions, as illustrated by Rajawat et al. Microbiome data analysis is expounded by Chauhan.

Many believe that the generation of omics data is most useful when it can be translated into three-dimensional structure. From the experimental perspective this requires careful studies of conformation and function in vitro. In this vein, several contributions in the book examine the experimental characterization of structures and interactions, such as that from Banerjee et al., who describe the use of circular dichroism to determine secondary and tertiary structure; Misra, who compares various types of fluorescence spectroscopy, and Kumar et al., who describe FTIR. Furthermore, detailed structures can be derived using NMR as described by Vazquez. A general chapter on microscopy is contributed by Baliyan and Kumar, covering TEM, SEM, and AFM. Thermodynamic aspects influence the states living systems find themselves in, and in this regard basic, underlying principles of isothermal titration calorimetry are explained by Srivastava and Yadav. The identification of phenotypes and mechanisms in cell-based arrays can be examined using flow cytometry, as described by Bajgelman.

Techniques such as those described in this book form an integral component of the armory required to translate molecular biological data into three-dimensional structure and interactions. Eventually, given enough supercomputing power, these data will be integrated into a fully three-dimensional computational model of the living cell, melding metabolism and machinery, with which the efficacy and safety of new drugs will be able to be tested in silico.

Sincerely,

*Jeremy C. Smith[1,2]*

[1]*Center for Molecular Biophysics, Oak Ridge National Laboratory, University of Tennessee, Knoxville, TN, United States,*
[2]*Department of Biochemistry and Cellular and Molecular Biology, University of Tennessee, Knoxville, TN, United States*

# Foreword

*The Data Processing Handbook for Complex Biological Sources* edited by Dr. Gauri Misra is a collection of articles representing almost all biophysical techniques currently in use for the analysis of biomolecular structure and function. Dr. Misra's extensive experience in editing such volumes is evident in all the articles included in the volume. Each chapter provides a comprehensive description of one of the methods and is authored by experts who have extensively used the technique and obtained important results as reflected in their publications. The chapters provide not only extensive references to published literature but also present the type of data obtained by each technique, methods of data analysis, expected results, and their biological interpretation. Although there are other volumes of similar content, the present volume is especially useful for several reasons. Students (and often teachers as well) of biology possess limited familiarity of physics and mathematics. The authors of the present volume have made substantial efforts to cover the theoretical and mathematical background necessary to understand the techniques. In the background of the theoretical basis of the methods, the authors present the nature of results obtained, data analysis protocols, their power, and limitations. If all the sophisticated instruments necessary for the experimental procedures covered in the volume are available in house, experiments can be repeated several times until satisfactory results are obtained. Due to the high cost of many of the instruments, very frequently it is necessary to use instruments available in other laboratories. The time available in such laboratories for outside users is limited and hence performing experiments successfully without repetitions is very essential. The authors of the various chapters have extensive experience in such efforts. Therefore, first time users of the techniques and those who do not possess in-depth expertise in physics and mathematics will find the volume invaluable. It is also a good reference volume to those who already possess expertise in one or more of these techniques.

*M.R.N. Murthy*
*Molecular Biophysics Unit, Indian Institute of Science, Bengaluru, India*

# Foreword

With new advanced technologies emerging, enormous biological data is generated. Analysis and organization of this data is important for gaining valuable information and to create new knowledge. The editor of this book, Gauri Misra, has a clear vision to bridge the gap between theoretical and practical aspects of biological data processing. Being a biologist she is able to provide a rational and broad outlook to experimental design and implementation. This is her third accomplishment as a book editor, which is certainly admirable for a young developing scientist. The emphasis of this book is on data processing approaches currently implied in a variety of biological systems from collecting raw data, organizing it, and interpreting processed data. No story is complete with information from a single technique or a scientific method. Amalgamation of various approaches is required to address the whole scientific problem. It appealed to me that the experts in different fields mastering related techniques came together to provide a detailed but clear explanation of various techniques are used in modern laboratories using examples from their own research work.

The most captivating part of this book is that it takes the reader with already a brief knowledge of the principles underlying various techniques to more advanced levels with excellent practical examples thus enabling this book to open to a wide audience in the scientific community. Also it is highly important for a student to be able to understand the concepts and the whole chain in biological data generation and processing to customize them for gaining solutions to their own research problems. This book certainly provides an excellent framework for both students and already experienced scientists. There is a lack of available data processing books that deal directly with the biological systems in the present market. Editor Gauri Misra has successfully filled the void. I hope this handbook will be a guiding light to scientists who try to use interdisciplinary approaches to validate their results. Albeit every person is not a master of all scientific domains and there is always a requirement for a quick handbook on how to analyze the data. This will be the first-hand resource that can be referred to by scientific personnel. The chapters include not only classical approaches such as circular dichroism, fluorescence microscopy, NMR, etc., but also the advances in these fields along with a chapter on more recent microbiome data analysis and high throughput sequencing. In my opinion, this book will be an asset for the scientific community that will help the readers to process and analyze the data generated from their own experiments.

Sincerely,

*Aatto Laaksonen[1,2]*

[1]*Physical Chemistry, Arrhenius Laboratory, Stockholm University, Stockholm, Sweden,*
[2]*Department of Materials and Environmental Chemistry, Stockholm University, Stockholm, Sweden*

# Preface

This is a continuation of the journey I embarked on with my first book on biophysics in 2017. A seed that was planted by the indefinite curiosity of the students I teach and the young scientists who try to amalgamate interdisciplinary approaches to enhance the worth of their research endeavors. These driving forces bear fruit in the form of the present book. There are various aspects of data analysis that have been emphasized in this book. The book is an interface that relates in-depth analysis of data generated from various biological platforms. The enormous amounts of data generated using several experimental approaches need to be interpreted in simple terms relating the biology and physics at the atomic level. The book is compiled with a focus on the target audience consisting of graduate students, young scientific professionals, and technicians from the industry sector who can use the data processing information for various techniques in this book to help them process their own data with ease.

The output files generated from various techniques are often not recognized by common user-friendly programs. When students read research papers they face a lot of difficulty in deriving the necessary information to process their raw data files. This book includes sample data files and explains the usage of equations and web servers cited in research articles to extract useful information from their own biological data. Details related to raw data files, data processing strategies, web-based sources relevant for data processing, and examples provide a unique perspective to the present effort. Unlike the other available books in the market on data analysis, which place more emphasis on computational and programming aspects, this book deals with biological aspects of data analysis underpinning the various experimental approaches used in modern biology.

The book is divided into 10 chapters dealing with different experimental techniques, namely mass spectroscopy, circular dichroism, fluorescence spectroscopy, high throughput sequencing, nuclear magnetic resonance, Fourier-transform infrared spectroscopy, microscopy, flow cytometry, isothermal titration calorimetry, and microbiome data analysis. Each chapter is a contribution from experts in the field thus enhancing the quality of the content and keeping the practical approach intact. It provides a quick guide for understanding experimental data. Each chapter begins with an introduction that guides the reader describing the structure, composition, and application of the chapter with the help of selective examples. The conclusion at the end of each chapter summarizes the key points and includes suggestions wherever required. However, an expert looking for a detailed explanation of anyone technique is advised to read further from suggested references given at the end of each chapter. I firmly believe that the book provides a valuable resource guide for students to plan and execute their own research experiments with a clear and simple understanding developed after reading this book.

*Gauri Misra, PhD*

# Acknowledgments

Any effort sees the light of the day with the blessings of the Almighty and one's parents. I acknowledge the hard work of all the contributing authors who have diligently tried to make each chapter worth reading. Thanks to all the students whose keen interest nurtured the desire to create this work. Some people leave an everlasting effect on your mind and shape your ideas. Two such people I wish to extend my gratitude are Dr. Shekhar C. Mande and Dr. M.R.N. Murthy.

My indebtedness to my mother, Mrs. Kamla Misra; my spiritual teacher, the late Bhaktivedanta Shri Narayan Swami Maharaj; my grandmother, the late Mrs. Shantidevi Misra; grandfather, the late Mr. Anand Swaroop Misra; and my school principal Sister Betty Teresa for their love, care, constant support, and freedom of choice to choose and shape my career in science and face the challenges of life, emerging as a strong person. Support and affection of friends has been a constant boost in this academic journey. I thank the reviewers for their valuable feedback that has certainly added weight to the enrichment of this project. The support from all the concerned Elsevier officials associated with this project is deeply acknowledged. With this I hope to progress in the future, nurturing new scientific ideas and creating many more masterpieces useful for the general scientific audience.

# Chapter | 1 |

# Mass spectroscopy

Jyotika Rajawat[1] and Gagandeep Jhingan[2]

[1]Molecular & Human Genetics Laboratory, Department of Zoology, University of Lucknow, Lucknow, India, [2]Valerian Chem Private Limited, New Delhi, India

## 1.1 Introduction

Mass spectrometry is an analytical technique that identifies biomolecules or proteins present in biological samples and also useful for studying protein—protein interactions. The information obtained can then be further employed to determine protein functions, functional relationships, and the cellular pathways regulated by the proteins. Protein—protein interactions yield insights on functional relationships including, enzyme-substrate, substrates, posttranslational modifying enzymes, protein—coactivator/corepressor, and host—pathogen interactions [1].

Mass spectroscopy—based proteomic workflow consists of the following steps:

1. Isolation of protein samples from biological samples and fractionation by 2D gel electrophoresis.
2. Fractionated proteins are tryptic digested and resulting peptides are further fragmented.
3. Peptides are injected in mass spectrometer (MS) and subjected to quantitative and qualitative analysis.
4. The dataset generated is analyzed by appropriate software for identification of the amino acid sequence and for protein quantification.
5. Protein identity is assigned based on database search.

## 1.1.1 Principle

Mass spectroscopy determines the characteristic fragments or ions that arise by organic molecule breakdown. The basic principle involves the bombarding of organic compound with a beam of electrons to generate positively charged ions. Further, the molecular ion is fragmented by utilizing energy of electrons to break the bonds and yield positively charged species or fragment ions. Positive ions or fragment ions formed are further accelerated and deflected in a circular path using magnetic field and then focused on the detector or photographic plate according to their mass and charge. Each ion represents a separate line on plate and recorded in form of peak intensity. Ion deflection is based on charge, mass, and velocity, ions separation is based on mass to charge ($m/z$) ratio and detection is proportional to abundance of ions.

## 1.1.2 Instrumentation

The first MS was designed by JJ Thompson in 1912 and was mainly used to study atomic weight of elements and to monitor the abundance of elemental isotopes. Later in the 1950s, with the advent of techniques to vaporize the organic compounds, it became possible to analyze biomolecules with a MS. Following are the main components of a MS as depicted in Fig. 1.1:

1. Ionization source
2. Analyzer
3. Detector
4. Data processor/digitizer

The analyzer and detector are kept in a vacuum so that the ions produced do not collide with air molecules and the ion trajectories are maintained at a particular speed. Once the sample is injected in the instrument through the inlet, it gets ionized in the ionization chamber. Ionized species are then extracted into the analyzer, which resolves the ions based upon their mass to charge ratio. Finally these ions are detected by detectors and relative

**Data Processing Handbook for Complex Biological Data Sources.** DOI: https://doi.org/10.1016/B978-0-12-816548-5.00001-0

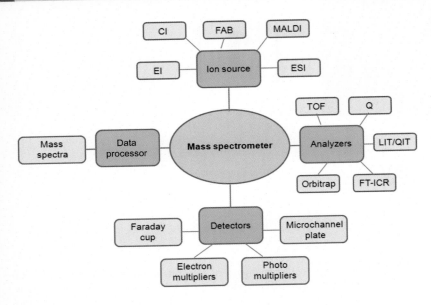

**Figure 1.1** Components of mass spectrometer.

abundance of each resolved ionic species is recorded in the form of mass spectra.

The samples can be injected into the instrument either by direct infusion or through multidimensional chromatographic separation. Analytes can be delivered in staggered manner with the use of chromatographic technique. Gas or liquid chromatography (LC) is commonly used for sample injection in MS. In gas chromatography (GC) the mobile phase used is gas such as hydrogen or helium, which injects the molecules of analytes in the column. The analyte can be effectively separated from the matrix by manipulating the column temperature. GC is limited to compounds that are volatile, heat stable, and lack polar functional groups. GC-MS is often used for comprehensive drug screening. LC is nowadays a preferred method for separation of samples as it is compatible to broad range of analytes and need less for derivatization than GC. The mobile phase uses aqueous and organic solvent in a particular ratio for altering the distribution of analytes and matrix. LC provides fast and effective separation of sample prior to MS analysis [2].

## 1.1.3 Types of ionization techniques

Different types of ionization methods are employed depending on the type of samples. Volatile samples are subjected to either electron or chemical ionization. Nonvolatile samples are ionized using any of the given techniques:

1. Fast atom bombardment
2. Matrix assisted laser desorption ionization (MALDI)
3. Electrospray

### 1.1.3.1 Electron ionization

The ionizing agent is high energetic electron emitted from a heated filament and accelerated due to potential difference of around 70 eV. Sample gets ionized by removal of an electron and generation of positively charged ion species.

$$M + e- = M^+ + 2e-$$

Charged radical cation further gets fragmented into positive ions. Electron ionization (EI) is widely used for volatile organic compounds like oils, hydrocarbons, etc.

### 1.1.3.2 Chemical ionization

The ionizing agent used is electron but the sample is indirectly ionized by reagent gases. Commonly used reagent gases are methane, isobutane, and ammonia. The vaporized sample along with excess reagent gas is injected into the spectrometer. Similar to electron ionization, electrons are generated with heated filament, which ionizes carrier gas to form primary ions, which further form secondary ions. These secondary ions then react with the sample and protonate it to produce protonated species. Chemical ionization (CI) is also used for volatile organic compounds.

### 1.1.3.3 Fast atom bombardment

Fast atom bombardment ionization technique is employed for less volatile large organic compounds like peptides, proteins, and carbohydrates. Sample is mixed with nonvolatile matrix usually glycerol or nitrobenzyl alcohol or crown ethers, and argon or xenon gas is then used to bombard the immobilized matrix. This bombardment generates charged sample ions, which are then focused into the analyzer.

### 1.1.3.4 Matrix assisted laser desorption ionization

It is similar to fast atom bombardment, wherein the sample is mixed with the matrix compound, which is capable of absorbing UV light. Excess fold of matrix is mixed with the sample using a solvent, which is then evaporated and finally cocrystallized analytes with crystal matrix are obtained. Short pulse of UV laser beam is then targeted on solid matrix resulting into ionization of samples. Due to energy absorption the matrix gets partially vaporized and hence carries analyte molecules in ionized form along with it in gas phase. During this process matrix and analytes exchange protons producing positively charged molecules [3].

### 1.1.3.5 Electrospray ionization

Electrospray is the softest ionization method, where it maintains the noncovalent interactions of a macromolecule in their gas phase. It is also termed as ion spray, sonic spray, or nanospray. Sample is passed through a high potential narrow electrical capillary resulting in an electrostatic spray composed of multiple charged droplets. Further these droplets pass under a heated capillary where solvent gets evaporated and gets exploded into microdroplets, wherein charged ionized analytes get separated from the microdrops that escape towards the analyzer. The major advantage of electrospray techniques is that ions acquire multiple charges depending on their molecular mass and hence peptides and proteins get successively protonated resulting into multiple peak spectra due to different $m/z$ ratios. It is useful for determining the molecular mass in low resolution spectrometers. Electrospray ionization (ESI) sensitivity is impoverished by miniaturized ESI sources, whereby the capillary opening size can be reduced and flow rate can be minimized [4].

### 1.1.3.6 Fragmentation

Molecular ions formed during the ionization process can further be fragmented upon bombardment with higher energy electron beam resulting the formation of fragments. Lowest ionization potential fragments retain the positive charge. Thus, a molecular ion upon fragmentation produces a radical and a cation; the latter is detected in MS. This chemical dissociation may be due to homolytic or heterolytic cleavage of single bond in molecular ion. Fragments resulted from fragmentation of molecular ions provide information about the structure of a molecule by analyzing the peaks generated in mass spectra. Fragment size and its frequency depend on the bond energy and the structure of the analyte molecule.

Factors important for fragmentation:

1. Stability of the bonds
2. Energy of the molecular ion and the fragments
3. Steric factors during rearrangement
4. Stevenson's rule: If the two fragments produced from a molecular ion compete for cation generation then the ion with lowest ionization energy will be formed. Also the largest fragment is lost in a branched radical cation.

$$AB ----A + B^+$$
$$Or \ AB ----A^+ + B$$

The fragmentation pattern depends on the functional group present in the compound.

Types of fragmentation:

1. Collision induced dissociation (CID): Molecules are allowed to collide with neutral gases leading to breakage of bonds and fragment ion formation. CID fragments are produced in triple quadrupole spectrometer.
2. Electron capture dissociation (ECD): A protonated molecule is subjected to low energy electrons resulting into production of an odd electron ion. ECD fragmented gas phase ions are used in tandem mass spectroscopic analysis.
3. Electron transfer dissociation (ETD): Fragmentation is done by transferring electrons to cations, as carried out for proteins and peptides.
4. Photo dissociation (PD): It is a chemical reaction where photons induce fragmentation of chemical compounds. Dissociation can be done using multiple infrared photons or carbon dioxide laser.
5. Surface induced dissociation (SID): Molecular ions in gas phase undergo fragmentation due to collision with a surface under high vacuum.
6. Charge remote fragmentation (CRF): A gas phase ion undergoes covalent bond breaking and cleaved bond is not next to the charge atom. Such fragmentation is used in tandem mass spectrometry.
7. High energy C-trap dissociation (HCD): This technique is applied in peptide modification analysis. HCD modification generates immonium ions.

## 1.1.4 Analyzers

Analyzers resolve and differentiate the ions based on their $m/z$ ratio and this separation is driven by electric or magnetic field. All mass resolved ions are focused on a single focal point. Desirable features of analyzer are:

1. Mass range is maximum permissible $m/z$ ratio.
2. Mass accuracy is expressed as parts per million (ppm) and determines how close is the measured mass to the accurate mass.
3. Resolution of MS is its ability to resolve molecular species with similar but distinct masses. Resolution is calculated by dividing the peak value of $m/z$ with its width at half maximum intensity.
4. Efficiency is the transmission multiplied by the duty cycle.
5. Sensitivity is the minimum concentration detected by the spectrometer.
6. Linear dynamic range.

The instruments can have two analyzers coupled together for mass analysis and such spectrometers are termed as tandem MSs, which we will discuss in the following section.

Types of analyzers:

1. Time of flight (TOF)
2. Quadrupole (Q)
3. Ion trap (LIT & QIT)
4. Fourier transform−ion cyclotron resistance (FT-ICR)
5. Orbitrap
6. Tandem MS

The analyzers vary in terms of mass range, sensitivity, resolution, dynamics, and performing tandem mass spectroscopy experiments. In the following section we will discuss these analyzers and tandem instruments.

### 1.1.4.1 Time of flight

It is the simplest mass analyzer, which consists of a flight tube placed in high vacuum. Ionized molecules fly through the flight tube with different velocities depending upon their mass, where mass is inversely proportional to the velocity. In a TOF analyzer, time of the flight to the detector is measured, where ions with lower mass will reach the detector earlier than higher mass ions. Flight time may vary from 1 to 50 µs. The TOF analyzer has several advantages: (1) all ions will reach the detector eventually depending on the TOF and (2) it can detect ions with very high mass range. The major drawback of TOF is the varying resolution due to different kinetic energies of ions entering the linear tube. To overcome this problem an electrostatic ion mirror, also known as a reflectron, is used

in TOF analyzers, which improves the resolution by narrowing down the kinetic energy distribution of ions due to deeper penetration of faster ions. Resolution is also improved by increasing the flight time with increase in tube length.

The majority of the TOF spectrometers are linked to the MALDI for source of ions. The current MALDI-TOF instruments can achieve peak mass resolution of 25,000 FWHM order or detect macromolecules with mass range of mega dalton. This technique is suitable for peptides, proteins, polynucleotides, lipids, and polysaccharides [3,4].

### 1.1.4.2 Quadrupole

It consists of four rods parallelly arranged in a hyperbolic section, whereby two opposite rods are connected to direct current (DC) and other two opposite rods are linked to radio frequency (RF) alternate current (AC). Opposite rods have DC voltage and other two have RF voltage 180 degrees out of phase thus generating an electrically oscillating field. When ions are pulsed from the ionization chamber toward the quadrupole, the positive ions will move toward the negative rod but due to changing polarity, the ions take up a complex trajectory. Thus trajectory of ions is regulated by varying the DC/Rf voltage. The ions within a narrow range of $m/z$ will take up stable trajectory in the quadrupole ultimately reaching the detector. Ions with improper trajectory collide with the rods and will not reach the detector. By manipulating the voltage and frequency, the ions with different $m/z$ ratios can be transmitted to the detector. In the RF mode where DC is zero, most ions acquire stable trajectory and pass through the quadrupole. In the scanning mode RF and DC simultaneously vary to permit ions with different $m/z$ ratio to pass through the quadrupole sequentially. The third mode is to stabilize the voltages thus permitting stable trajectory for ions with predetermined $m/z$ ratio. Major disadvantages of the quadrupole are the limited mass range up to 4000 Da, lower resolution and decreased ion transmission with increased $m/z$ ratio. The quadrupole's ability for MS/MS analysis can be enhanced by attaching another analyzer to it. Quadrupole when coupled with ESI ion source determines the molecular weight of proteins and nucleic acids.

### 1.1.4.3 Ion trap analyzer

As the name indicates, ions are trapped in a long tube and can be accumulated during the time of operations in the instrument. The conventional ion trap analyzers are quadrupole ion trap (QIT) or 3D IT, comprising of a ring electrode and two endcap electrodes. These three electrodes are arranged in a manner to create a cavity for trapping ions and analyzing them. RF voltage and electrodes are designed in a way to generate stable trajectories of ions, which remain entrapped in instrument for seconds to hours. Increasing the RF voltage destabilizes the ion trajectory and ions are ejected to the detector. Besides MALDI-TOF, QIT is the most common MS instrument for proteomic studies. When coupled to ESI source QIT has high sensitivity for peptide analysis and is popularly used in proteomics. Generally with QIT both full scan and product scan are carried out at high scan speed and low resolution. Nevertheless good resolution can be obtained at lower scan speed when only short $m/z$ is to be considered. A disadvantage of QIT mass analyzers is low resolution, limited ion trapping capacity, and low sensitivity due to space charging effect resulting in lack of accuracy in mass measurement [3,5].

Linear ion trap (LIT) mass analyzers have been developed recently to overcome the drawbacks of QIT, expanding the sensitivity and dynamic range of the ion trapping technique. LIT or 2D ITs have higher ion trapping capacity where 50 times more number of ions can be trapped, and have higher trapping efficiency of one order higher magnitude than QIT. LIT is comprised of quadrupole rods with RF voltages and end with DC arranged to form a cavity for trapping ions. These instruments have higher resolution with incorporation of optional slow scanning mode. Triple quadrupole instruments with a second analyzer substituted with LIT results in a Q-Q-LIT MS/MS instrument, which has higher scanning speed and sensitivity and is also capable of posttranslational modifications like phosphorylation.

### 1.1.4.4 Fourier transform—ion cyclotron resistance

The ICR MS is composed of four electrodes placed in a magnetic field such that a Penning trap is created where the electric field is perpendicular to the magnetic field. The ions get trapped in the Penning trap and oscillate with cyclotron frequency. Cyclotron frequency is inversely proportional to $m/z$ ratio and directly proportional to magnetic field intensity. The signal is detected in form of image current on the detector plates and this signal is termed as free induction decay (FID), which has frequency equal to the cyclotron frequency of ions. Useful information or data is extracted from this signal by using a mathematical equation known as Fourier transform (FT), thus yielding mass spectra. FT-ICR can detect molecules with low ppm to sub ppm range thus improving on mass accuracy measurement. Currently it is the only MS with highest resolution of approximately $10^6$ values and thus intact undigested proteins can be studied. Higher resolution also corresponds to increased peak capacity and hence more signals can be detected.

### 1.1.4.5 Orbitrap analyzer

Orbitrap analyzer works on the principle where ion separation occurs in an oscillating electric field. Orbitrap mass analyzer has inherited key features of previous analyzers, like electrostatic field adapted from TOF, ion trapping within the electrodes from ion trap analyzers, and image current measurement from FT-ICR. Mass accuracy measurement and resolution of orbitrap analyzer is similar to FT-ICR but overcomes the limitation of expensive superconducting magnet. Orbitrap analyzer is comprised of three electrodes, where the outer two electrodes face each other in the form of cups and are separated by a dielectric central ring. The central electrode is spindle shaped and holds the trap and aligns it towards the dielectric end spacers. The radial electric field between the outer and central electrode and the centrifugal force pushes the ion to take up harmonic axial oscillations. Outer electrodes serve as receiver plates, which detect axial oscillations in form of image current detection, and then further signal is transformed similar to FT-ICR into mass spectra [6].

### 1.1.4.6 Tandem mass spectrometer

When two or more than two mass analyzers are connected in sequence to reveal in-depth information about the molecular structure of analyte then such instrument is termed as tandem MS or MS/MS instrument. Mass analysis of analyte is carried out in two consecutive stages, wherein the first analyzer performs ion isolation, then fragmentation induced by CID in collision cell, and finally followed by separation of fragmented ions based on $m/z$ ratio in the final analyzer. Thus there are two scanning modes: the first is parent scan mode in the first analyzer, which detects the precursor ion of interest; and the second is product ion scan or MS/MS scan or fragmentation scan where fragmented ion is detected. Tandem MS can be of two types:

1. Tandem-in-space
2. Tandem-in-time

Tandem-in-space are MS/MS instruments where analysis is performed in different space, that is, mass analyzers. Triple quadrupole (QqQ or TQ) is the common tandem-in-space MS/MS instrument where Q denotes two quadrupole analyzers and q denotes the collision cell. These quadrupoles mostly operate in rf mode, specifically the collision cell q. The third analyzer in TQ when replaced with other mass analyzers results in hybrid tandem MS like Q-TOF and Q-LIT MSs. Q-TOF combines the positive features of two analyzers: the scanning power of quadrupole and resolving capacity of TOF analyzer. The advantage with Q-TOF analyzer is simultaneous monitoring of several ions, in contrast to single ion checking in TQ. Thus Q-TOF provides higher sensitivity and resolution of one order higher magnitude than the TQ instrument. Tandem MS is mostly coupled to ESI or MADLI for ionization source. Such instruments provide good quality of fragmentation spectra resulting in protein identification. Tandem MS instruments are very useful when protein cannot be identified by peptide mass fingerprinting (PMF). QTRAP spectrometer has been developed for especially for analysis of posttranslational modifications in proteins.

Tandem-in-time are trapping MS instruments that eject all ions except for one $m/z$ that gets fragmented in same space. Ion trap quadrupole and FT-ICR MS is such an instrument that performs multiple MS experiments (MS") in a single analyzer.

## 1.1.5 Detectors

Basically there are four types of detectors used in MSs:

1. Faraday cup
2. Electron multiplier
3. Photomultiplier
4. Microchannel plate

### 1.1.5.1 Faraday cup

The dynode consists of secondary emitting materials such as GaP, BeO, and CsSb. The principle of the Faraday cup is that when incident ions or ion beam strikes the dynode, electrons are emitted that induce current, which is amplified and recorded. Such detectors are suitable for isotope analysis.

### 1.1.5.2 Electron multipliers

Dynodes are made up of copper-beryllium. Electron multipliers are especially useful when both positive and negative ions have to be detected. The principle involves the emission of electron from first dynode upon transducing initial ion current, which is then focused to the next dynode and so on. This cascade of current is finally amplified a million times and recorded.

### 1.1.5.3 Photomultipliers

The dynode is made up of a substance called the scintillator, which emits photons. Light emitted from the dynode is converted into electric current by a photomultiplier tube and current is recorded. Photomultiplier detectors are suitable to study metastable ions.

### 1.1.5.4 Microchannel plate

It comprises of an array of miniature electron multipliers, which act as a continuous dynode. The channels are comprised of a special glass with coating of high resistance semiconductor and these semiconductors function as electron multipliers. The multiple electrons emitted from the channel produce a pulse equivalent to $10^6$ electrons, which is emitted from a small spot at the rear end of microchannel plate and generated as the output and ultimately recorded.

The overall workflow of a MS is depicted in Fig. 1.2.

## 1.1.6 Mass interpretation

Based on mass spectra obtained from peptide analysis, protein identification is done by mainly three approaches:

1. Peptide mass fingerprinting (PMF)
2. Peptide fragmentation fingerprinting (PFF)
3. De novo peptide sequencing

Before studying the details of these approaches for peptide identification we will have a brief look into the peaks obtained in mass spectrum and the fragmentation rules.

### 1.1.6.1 Types of peaks

1. Molecular ion peak: It represents the molecular ion produced when sample is bombarded with $9-15$ eV energy electrons.
2. Fragment ion peak: Molecular ions, when further bombarded with high energy electron of up to 70 eV, result in fragment ions, which correspond to fragment ion peak.
3. Rearrangement ion peak: it is the representation of recombination of fragment ions.
4. Metastable ion peak: decomposition of analyte between source and magnetic analyzer results in formation of metastable ions, which correspond to broad peaks known as metastable ion peaks.
5. Multicharged ion: Some ions are formed with more than one charge and are represented as multicharge ion peaks.
6. Base peak: the largest and most intense peak in the mass spectrum is known as the base peak.
7. Negative ion peak: capture of electron by a molecule during electron bombardment results in negative ion formation, which forms negative ion peak in the spectrum.

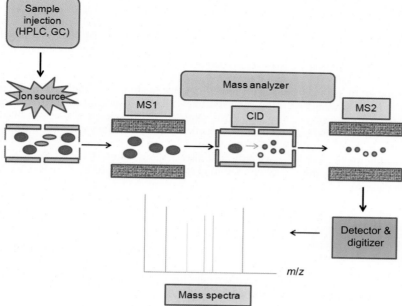

**Figure 1.2** Outline depicting the workflow of mass spectrometer.

## 1.1.6.2 Fragmentation rules

There are nine rules to analyze fragmentation peaks in the mass spectrum.

Rule 1: Increase in degree of branching results in decrease in M + peak height.

Rule 2: Increase in molecular weight decreases M + peak height.

Rule 3: M + peak is stabilized by aromatic rings, cyclic structures, and double bonds.

Rule 4: Allylic cleavage favored by double bond results in resonance stabilized cation.

Rule 5: Cleavage occurs at alkyl substituted carbon resulting in carbocation formation,

Rule 6: Alkyl side chain in saturated ring is removed while retrodiels−alder reaction occurs in unsaturated rings.

Rule 7: Aromatic compounds are preferably cleaved at β-bond resulting in resonance stabilized benzyl ion.

Rule 8: C−C bonds next to hetero atom (O, NH, or S) are cleaved frequently, producing charged hetero atom, which is stabilized by resonance.

Rule 9: There has been elimination of small neutral molecules during cleavage of bonds along with rearrangements of atoms. This is known as McLafferty rearrangement.

## 1.1.6.3 Peptide mass fingerprinting

PMF is the most commonly used strategy to identify proteins and protein expression pattern in biological samples. The mass profile obtained after MALDI-TOF MS is compared with the predicted mass values obtained from the in silico digestion of all proteins present in the database. Common search programs used in PMF are MASCOT, PROFOUND, MSFIT, and ALDENTE, which can be easily accessed through the Internet. Four or five peptide mass matches are sufficient to identify the correct protein hit, covering 15% of the sequence and mass accuracy better than 30 ppm. Successful protein identification depends on the following factors:

1. Peak mass accuracy
2. Assigned and unassigned peaks relation in the spectrum
3. Database size for analysis

   *Drawbacks to PMF:*

1. Masses of the experimental peak might not match with the peptide masses of the theoretical proteins
2. Not all peptide masses are generated theoretically upon in silico digestion in the database [3,7].

## 1.1.6.4 Peptide fragmentation fingerprinting

This approach uses low energy collision induced fragmentation (CID) for peptide ions fragmentation in gas phase in MS/MS instrument. Such fragmentation occurs mainly at peptide amide bonds and yield two types of fragments:

$\beta$ ion, which preserves N-terminus, and $\gamma$ ion with C-terminus. The fragmentation spectra obtained is compared with the search programs containing the predicted spectra of peptides after in silico digestion of all available proteins in the database. MASCOT and SEQUEST are the programs used to analyze the CID fragmentation spectra and identify the peptides and proteins. A single peptide can identify a protein match but a greater number of peptide matches with one protein increases the sequence coverage and increased probability of correct match. Unfortunately the disadvantage for peptide fragmentation spectra analysis is that this approach is not applicable if there is any mass difference because of certain modifications or if there is no match to the peptide in the database as the protein is not reported in that particular database.

### 1.1.6.5 De novo peptide sequencing

De novo peptide sequencing is reconstitution of the peptide sequence from the fragmentation spectra based on peptide fragmentation rules. It is a straightforward approach for protein identification when protein identification becomes elusive. This task was traditionally done by Edman's sequencing but with the advent of the latest technologies and bioinformatics tools, the data produced is searched for protein homology using BLAST tool.

In our next section we have discussed raw data processing, focusing mainly on proteins.

## 1.2 Raw data

The last 10 years has seen a significant growth in proteomics study. The credit goes to the development of compelling protein profiling technologies, which has led to research in protein biomarker discoveries of various illness and diseases thereby making a remarkable improvement in the diagnostic and therapeutic field. Proteomics, in a broader sense, involves study of a specific proteome, validation of information about protein abundances, its modifications along with protein–protein interactions, and its interacting partners. These kinds of studies come under discovery proteomics, which involves two kinds of approaches, namely, top-down and bottom-up approaches. The former involves protein characterization by MS without undergoing proteolysis, thus retaining mass ranges to a greater extent in the case of PTMs, detection of "native" molecular mass of proteins, and greater sequence coverage. Unlike top-down, bottom-up approach involves proteolytic digestion of crude protein samples before running in MS. The profiles of these runs are acquired simultaneously by a PC coupled with MS.

The method of data acquisition can be classified into data-dependent acquisition and data-independent acquisition. Data-dependent acquisition involves selection of certain peptides to be fragmented at the MS/MS level based on specified standards. Data-independent acquisition involves selection of all peptides within a certain mass range at MS/MS level. After data acquisition, raw files are taken for protein/peptide identification and quantification using software, like TPP, which employs search engines like SEQUEST, OMSSA, Mascot, etc., and the quantified proteins are statistically analyzed using statistical software like R, Perseus, etc.

We will now discuss the nuances of Nano-ESI LC-MS/MS with Orbitrap technology followed by data-dependent acquisition, Trans-Proteomic Pipeline (open-source for protein/peptide identification and quantification), label-free quantification, and various Annotation Software tools.

*Nano-ESI LC-MS/MS*: It is a powerful setup that comprises of nanoscale LC paired with tandem mass spectrometry. unlike the traditional LC-MS/MS techniques, this setup provides high sensitivity and nanoflow UHPLC performance.

### 1.2.1 NanoLC principle

The term "nanoLC" refers to flow rates in the $nL\,min^{-1}$ range (as opposed to high $\mu L\,min^{-1}$ to low $mL\,min^{-1}$ range for regular HPLC). NanoSpray mass spectrometry with nanoLC flow rate and concentration-sensitive detectors usually separates a minute amount of samples at microscale. A typical nanoLC/MS/MS setup will be a 75-$\mu m$ I.D. column at a flow rate of $200-350\,nL\,min^{-1}$. In recent times, gel-free analytical approaches based on LC and nanoLC separations have been developed, leading to faster and more extensive proteomics-based studies than the classical 2D gel electrophoresis approach.

It is here that the biological samples, which are proteolytically digested using an enzyme and broken down to peptides, are injected into liquid chromatography (HPLC), strategically forming a gradient along with the injected sample and getting eluted into the MS.

### 1.2.2 Gradient

Gradient method is employed for those samples that have wide "$k$" range and hence cannot be separated using isocratic methods. Gradient in reversed-phase HPLC is specified by three essential parameters: initial %B, final %B, and gradient time ($t_G$) over which the transition in eluotropic strength will be achieved. The most frequent questions in gradient HPLC are asked about how to

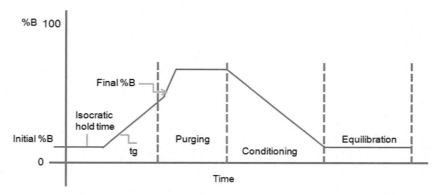

**Figure 1.3** Diagrammatic representation of different steps involved in the gradient elution.

decide upon the optimum values for these three parameters, and it is conventional to begin method development with a "scouting gradient" as the eluotropic strength increases from a low to high value over a set period, typically 5%−95% B over 10 or 20 minutes. The optimum operating conditions can then be assessed by the analyte elution behavior during this scan (Fig. 1.3).

After forming a gradient, the sample reaches the MS and undergoes three main stages:

1. *Ionization*: An ESI source acts as a junction to couple LC with MS. The sample or analyte mixed with solvent, in its liquid state pushes through a capillary to form a jet of aerosol (miniature droplets of 10 μm), which is further nebulized by nebulizer gas like nitrogen. During nebulization, the solvent evaporates and the analyte ions happen to undergo fission, and these ions enter the MS to be separated by their $m/z$ ratio. For nanoLC, nanospray ionization (NSI) source is used where the sample elutes from a nano-ESI emitter.
2. *MS1 scan*, where $m/z$ ratios are measured and abundances are calculated. Here, a specific $m/z$ range is selected from the precursor ion.
3. *MS2 scan*, where fragmentation of ions takes place and $m/z$ ratios are measured and detected, yielding fragment ion spectra. Here, product ions are captured for detection.

## 1.2.3 Orbitrap technology

LTQ Orbitrap and LTQ Orbitrap Velos represent a consolidation of two mass analyzers, a LIT and an Orbitrap mass analyzer. It offers up to three fragmentation approaches: CID or ETD in the bifold LIT, or HCD in its devoted collision cell [8]. The fragment ions generated by CID or ETD are normally analyzed in the ion trap at low resolution, along with parallel acquisition of the MS transient. Nonetheless, they can also be conveyed to the C-Trap and registered in the Orbitrap analyzer. HCD fragments are always analyzed in the Orbitrap analyzer [9].

## 1.2.4 Data-dependent acquisition

This is the basic mode of data acquisition used in LC-MS/MS related experiments across various biological samples, for protein identification and relative quantification. Here peaks are identified based on preset values from a survey scan and are further taken in for MS/MS analysis.

After MS1 survey scan, we get main base peak with maximum intensity, which corresponds to the protein of interest. As shown in Fig. 1.4, the peak with $m/z$ value of 1424.6 is the base peak (mother peak). Peptide masses are determined from the LC-MS spectrum, and further, for MS/MS analysis or MS2 scan, the precursor ion of desired peak (each peptide) is subjected to fragmentation. Ions with certain $m/z$ values are only further subjected to fragmentation. There is a list of $m/z$ values in-built in the software on the basis of which certain ion peaks will be eliminated from further analysis. Fragment ion spectra are obtained resulting in daughter peaks, where the last peak corresponds to $m/z$ value of base peak 1424.6, as seen in Fig. 1.5. Peptide fragment ion mass is determined for each peak obtained from MS/MS spectrum. Peptide fragment ion mass is then coupled with the peptide mass of LC-MS spectrum and subjected to peptide scoring through database search. Protein is then identified from the correctly identified peptide scoring.

In this chapter we have also discussed interpreting the mass spectrum of phosphoprotein. Below is an example of mass spectrum of a nonphosphopeptide and a phosphopeptide. Peptide mixture of Mycobacterium kinase

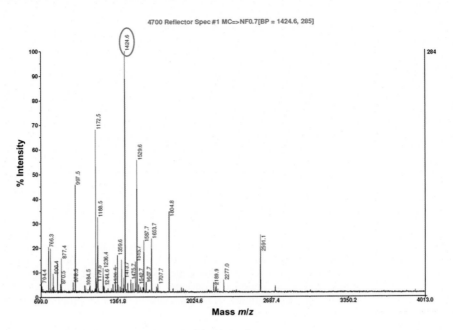

**Figure 1.4** Mother peak obtained after mass spectrometry (MS1 scan).

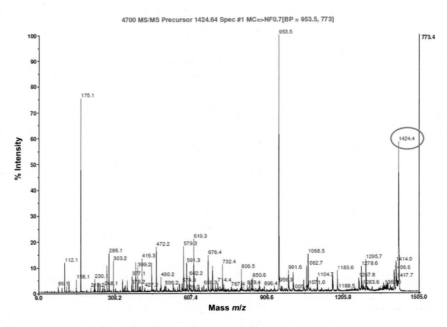

**Figure 1.5** Daughter peak obtained after MS/MS fragmentation analysis.

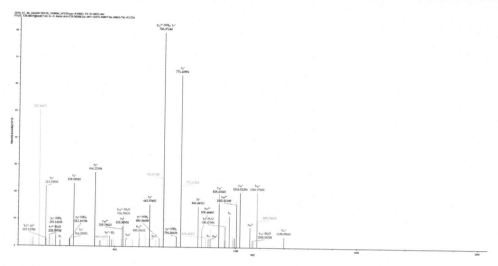

**Figure 1.6** MS/MS spectrum of nonphosphopeptide during HCD fragmentation.

**Peptide Summary**

Sequence: VLTDAERTSLLSSAAGNLSGPR, Charge: +3, Monoisotopic m/z: 739.06073 Da (-0.02 mmu/-0.03 ppm), MH+: 2215.16764 Da, RT: 53.2652 min,
Identified with: MS Amanda 2.0 (v2.0); Amanda Score:210.76, Percolator q-Value:0, Percolator PEP:0.000204, Ions matched by search engine: 26/61
Fragment match tolerance used for search: 0.5 Da
Fragments used for search: b; y

**Fragment Matches**

Value Type: Theo. Mass [Da] ▾

Ion Series | Neutral Losses | Precursor Ions

| #1 | b⁺ | b²⁺ | b³⁺ | Seq. | y⁺ | y²⁺ | y³⁺ | #2 |
|----|-----|------|------|------|-----|------|------|----|
| 1 | 100.07569 | 50.54148 | 34.03008 | V | | | | 22 |
| 2 | 213.15975 | 107.08352 | 71.72477 | L | 2116.09928 | 1058.55328 | 706.03795 | 21 |
| 3 | 314.20743 | 157.60735 | 105.40733 | T | 2003.01522 | 1002.01125 | 668.34326 | 20 |
| 4 | 429.23438 | 215.12083 | 143.74964 | D | 1901.96754 | 951.48741 | 634.66070 | 19 |
| 5 | 500.27149 | 250.63938 | 167.42868 | A | 1786.94060 | 893.97394 | 596.31838 | 18 |
| 6 | 629.31408 | 315.16068 | 210.44288 | E | 1715.90349 | 858.45538 | 572.63935 | 17 |
| 7 | 785.41519 | 393.21124 | 262.47658 | R | 1586.86089 | 793.93408 | 529.62515 | 16 |
| 8 | 886.46287 | 443.73507 | 296.15914 | T | 1430.75978 | 715.88353 | 477.59145 | 15 |
| 9 | 973.49490 | 487.25109 | 325.16982 | S | 1329.71210 | 665.35969 | 443.90889 | 14 |
| 10 | 1086.57896 | 543.79312 | 362.86451 | L | 1242.68008 | 621.84368 | 414.89821 | 13 |
| 11 | 1199.66303 | 600.33515 | 400.55919 | L | 1129.59601 | 565.30164 | 377.20352 | 12 |
| 12 | 1286.69506 | 643.85117 | 429.56987 | S | 1016.51195 | 508.75961 | 339.50883 | 11 |
| 13 | 1373.72708 | 687.36718 | 458.58055 | S | 929.47992 | 465.24360 | 310.49816 | 10 |
| 14 | 1444.76420 | 722.88574 | 482.25958 | A | 842.44789 | 421.72758 | 281.48748 | 9 |
| 15 | 1515.80131 | 758.40429 | 505.93862 | A | 771.41078 | 386.20903 | 257.80844 | 8 |
| 16 | 1572.82278 | 786.91503 | 524.94578 | G | 700.37366 | 350.69047 | 234.12941 | 7 |
| 17 | 1686.86570 | 843.93649 | 562.96009 | N | 643.35220 | 322.17974 | 215.12225 | 6 |
| 18 | 1799.94977 | 900.47852 | 600.65477 | L | 529.30927 | 265.15827 | 177.10794 | 5 |
| 19 | 1886.98180 | 943.99454 | 629.66545 | S | 416.22521 | 208.61624 | 139.41325 | 4 |
| 20 | 1944.00326 | 972.50527 | 648.67260 | G | 329.19318 | 165.10023 | 110.40258 | 3 |
| 21 | 2041.05602 | 1021.03165 | 681.02353 | P | 272.17172 | 136.58950 | 91.39542 | 2 |
| 22 | | | | R | 175.11895 | 88.06311 | 59.04450 | 1 |

**Figure 1.7** Peptide (nonphosphopeptide) summary and fragment matches using database.

was resolved and MS data acquired. Figs. 1.6 and 1.8 represent the mass spectrum of nonphosphopeptide and phosphopeptide respectively. MS data was acquired using a data-dependent top 10 method dynamically choosing the most abundant precursor ions from the survey scan.

The raw file was then analyzed with Proteome Discoverer 1.8 (Thermo) against the Uniprot *Mycobacterium Tuberculosis* database. For MS Amanda search, the precursor and fragment mass tolerances were set at 10 ppm and 0.5 Da, respectively. The protease used to generate

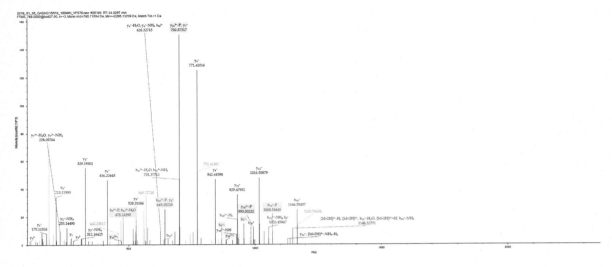

**Figure 1.8** MS/MS spectrum of a phosphopeptide.

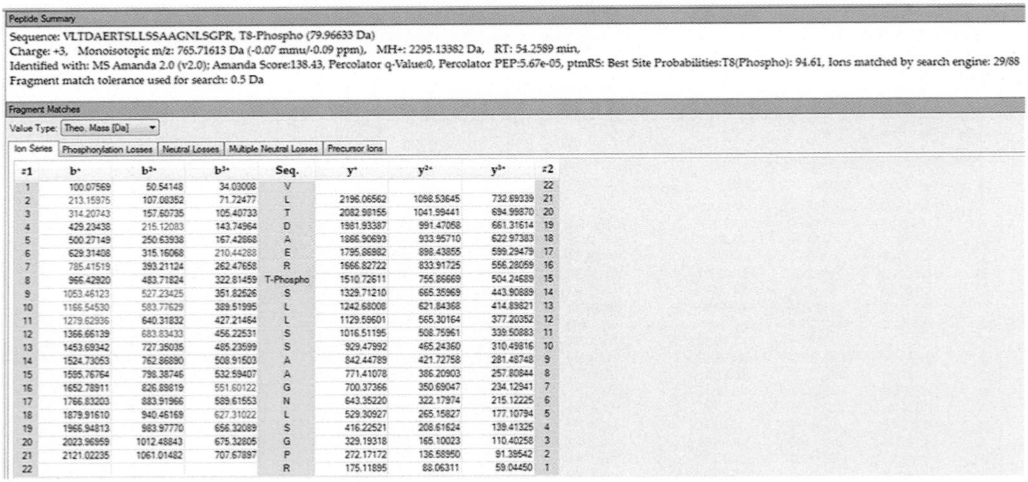

**Figure 1.9** Peptide summary and fragment matches of a phosphopeptide using database.

peptides, that is, enzyme specificity was set for trypsin/P (cleavage at the C-terminus of "K/R: unless followed by P") along with maximum missed cleavages value of two. Carbamidomethyl on cysteine as fixed modification and phosphorylation at S, T, Y, oxidation of methionine and N-terminal acetylation were considered as variable modifications for database search. Both peptide spectrum match and protein false discovery rate were set to 0.01 FDR (Fig. 1.7).

Fig. 1.8 represents MS/MS spectrum of precursor $m/z$ 765.71613 (+3) and $MH^+$ 2295.15 Da, of the semitryptic phosphopeptide VLTDAER (pT) SLLSSAAGNLSGPR, where pT indicates the site of phosphorylation. The unambiguous location of the intact phosphate group was determined by the presence of the "B" and "Y" ion series containing $B_{9-12}$, $B_{16}$, $B_{18}$ and $Y_{15}$, $Y_{18}$, $Y_{20}$, $Z_{18}$ (Fig. 1.9).

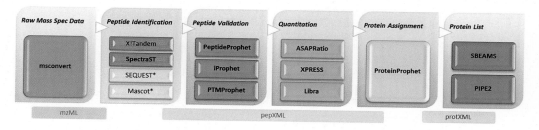

**Figure 1.10** Trans-Proteomic Pipeline workflow.
*http://www.socr.umich.edu/CSCD/html/Cores/Macore2/Macore2_2015.html.*

## 1.3 Data processing

### 1.3.1 Trans-proteomic pipeline

This online software provides a global platform to perform various tasks from MS/MS raw file conversion using MSConvert GUI to peptide and protein-level identification and quantitation. It is open source and provides tools that can be run individually or can be used as a conduit to perform various tasks, and are easy to follow for analysis and validation of the dataset (Fig. 1.10). http://tools.proteomecenter.org/wiki/index.php?title = TPP:5.1.0_Release_Notes

### 1.3.2 Installation

A newer version of TPP 5.1.0 is available for downloading, for Windows users, at https://sourceforge.net/projects/sashimi/files/. This version unlike one of its older versions, namely TPP 4.8.1, does not require dependencies. As for the Windows users, the older version of TPP has to be uninstalled to install the newer version. TPP is available for OSX users as well, but is not official yet. For Linux users, follow the link for installation: http://tools.proteomecenter.org/wiki/index.php?title = TPP:5.1_Installation

*Steps for installation:*

- Install the latest version of ActivePerl, preferably 5.9 or higher (www.activestate.com/activeperl/downloads)
- Download the latest version of TPP and Run the .exe file.
- Accept the license agreement and select a location to install TPP and its components. "C:\TPP\" is set as default and can be changed according to preference.

- When asked for the port number, set default 10401.
- Select the driver location for the data.
- Choose all extra components to install:
  - Apache web server
  - ProteoWizard
  - Download and install dependencies for ProteoWizard
  - Strawberry Perl
- Reboot after completing TPP Setup Wizard.

*The whole installation process takes around 10−30 minutes. After installation, TPP Petunia interface will be available on the Desktop or Start menu. Launch TPP icon and log in when the browser window opens. For detailed tutorial on TPP, follow the link: http://tools.proteomecenter.org/wiki/index.php?title = TPP_Demo2009

The following details will shed some light on the workflow components and tools of TPP and its analysis. Also, to get a better idea of how to perform the analysis, we have included a demo tutorial analysis, which contains the raw files and files of other formats derived out of TPP tools (****data from sourceforgenet*****). Log in to TPP GUI and add the mzML files CONTROL.mzML and TREATED.mzML.

### 1.3.3 Raw file conversion

File format conversion and further analysis of data can be well understood by the flow diagram (Fig. 1.11).

Depending on the instrument type, the formats of raw files generated from MSs have to be first converted to an open format mzXML or mzML. TPP uses the MSConvert tool for this purpose.

*Tutorial:* Add the raw files downloaded from sourceforgenet to the subdirectory, which directs to the drive where the analysis folder will be located.

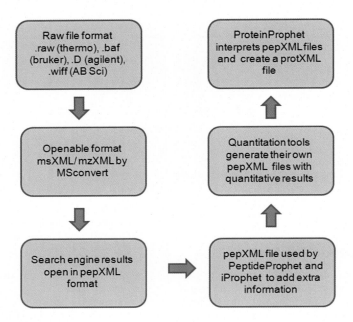

**Figure 1.11** Schematic workflow of "Downstream processing" file format conversion.

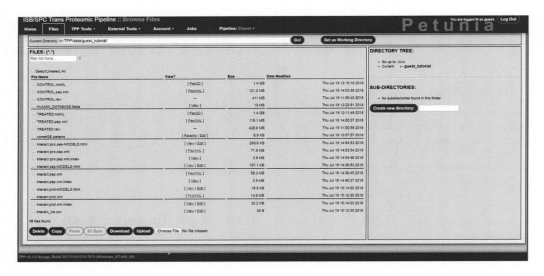

## 1.3.4 Search engine support

The MS/MS raw files can be processed initially by matching spectra (hypothetical spectra estimated from sequence database) with extracted peptide sequences. TPP supports search engines like X!Tandem, Mascot, OMSSA, SEQUEST, ProbID, and Phenyx, of which only X!Tandem is available with TPP. The rest need to be installed individually. The raw files can also be processed by matching spectra using spectral libraries (existing spectral library), which is the preferred method to obtain a good result, in comparison with hypothetical spectra match.

Some of the open source database search algorithms are given below:

1. Andromeda (part of MaxQuant)
2. Comet
3. Tide (rewrite of Crux)
4. Greylag
5. InsPecT
6. MassMatrix
7. MassWiz
8. MS-GF +
9. MyriMatch
10. OMSSA
11. pFind
12. ProbID
13. ProLuCID
14. Protein Prospector
15. SIMS
16. SimTandem
17. SQID
18. X!Tandem

*Tutorial:* For this tutorial, we have used Comet Search Engine coupled with Human database, which has been provided in the directory. While carrying out search, the process will look like this:

## 1.3.5 PeptideProphet for corroboration

PeptideProphet is a software component developed to perform validation of peptides allocated to MS/MS spectra. One of its features includes employing estimation maximization algorithm to calculate PSM probability. Additionally, it can also provide information on mass deviation, missed cleavages, and retention time. In other words, it assigns probabilities that peptides assigned to MS/MS spectra are true. This option can be set to avoid information on decoy for separate evaluation at the end result. The resulting pepXML format file can be analyzed with PepXMLViewer Application. http://peptideprophet. sourceforge.net/

*Tutorial:* Go to "Analyze Peptides" option in TPP tools to add the mzML files and set the default options available for PeptideProphet and then run analysis.

Final result of PeptideProphet can be viewed using PepXML viewer as shown below.

## 1.3.6 Pep3D to view chromatogram

To view the quality of sample run in MS, Pep3D allows the user to view chromatograms from various LC runs.

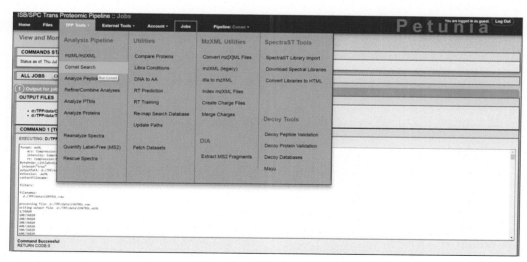

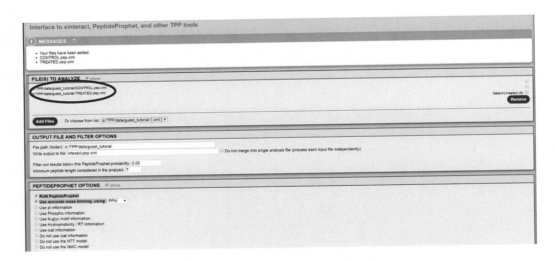

*Tutorial:* PepXML Viewer has a feature to view the chromatogram of sample. Go to Other actions menu, the last option located in the viewer and click on "Generate 3D" option with defaults set as it is.

## 1.3.7 iProphet for peptide-level corroboration

iProphet provides the rightly identified peptide sequences with high posterior probability and FDR by assimilating same documented peptide sequences of the same data searched via various search engines. It also couples the information collected about the same peptide spectra across various experiments, other diverse spectra, precursor ion charge, and modified states, thereby creating a multifaceted model to corroborate the peptide. This further increases the confidence at which the peptide is being scored with high confidence. It is because of such profound analysis done at peptide level that this software tool is considered to be more effective than PeptideProphet or Percolator [10].

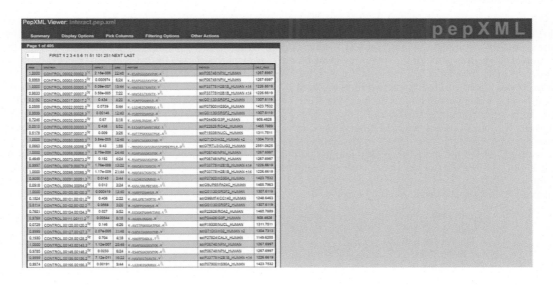

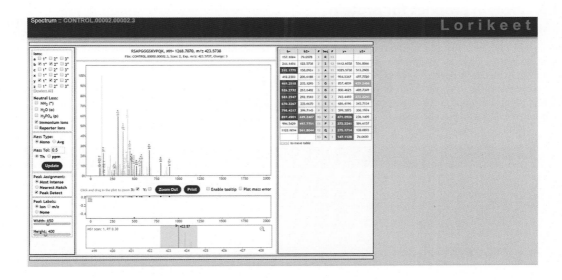

*Tutorial:* To further validate the peptides statistically, go to "Refine/Combine Analysis" option in TPP tools menu and uncheck all other Search parameters except that of iProphet and upload the pepXML files. As a result a detailed and much refined group of peptides are obtained in this process. The result can be viewed using PepXML viewer. As PSMs are calculated for each peptide, the quality of individual ion spectra can be viewed when clicking on "ions" of each peptide, as shown below:

## 1.3.8 XPRESS/ASAP ratio for peptide quantitation

*XPRESS* tool quantifies ratios of differentially labeled proteins computing relative abundance of proteins. The precursor ions are remodeled from light and heavy elution profiles, where the area of each peak is found. It also cites labeled residues and determines mass difference between two isotope labels.

*ASAP Ratio* (Automated Statistical Analysis on Protein Ratio) on the other hand is a tool that uses a refined way to measure and accumulate peptide ions at the peptide and protein level, thus generating a list of proteins based on their relative abundance. It also sets another list of proteins that will show PTMs. These two tools are used in case of heavy- and light-labeling techniques like ICAT or SILAC. https://omictools.com/xpress-tool

However in the case quantitation to be performed from isobaric labeling techniques like iTRAQ, relative abundances have to be calculated using relative peak intensities of the reporter ion. Another tool called *Libra* performs quantitation for analysis of this kind. Some of the open source MS/MS peptide quantitation software is given below:

1. MassChroQ
2. MaxQuant
3. OpenMS/TOPP
4. ProtMax
5. Skyline
6. BACIQ

## 1.3.9 Protein prophet for protein interpretation and corroboration

Protein Prophet, like PeptideProphet, confirms that the proteins based on peptides assigned to MS/MS spectra are true. It mainly directs to less-likely "single hit" proteins and their corresponding peptides and looks upon the peptide probabilities to group the proteins accordingly. Also, it helps the peptides to be allocated to their respective

proteins in case of multiple databases used. The result is produced in protXML files, which are in HTML format.

*Tutorial:* After deriving result from iProphet in pepXML format, this is used further to group proteins corresponding to peptides to which they belong. For this, go to "Analyze Proteins" option in the TPP tools menu and add pepXML files to be analyzed. Check on the "Input is from iProphet" option in ProteinProphet and run Protein Analysis. The result will be of protXML format, which can be viewed using View option. The final result looks like this:

**After deriving protXML file, it can be further used to perform Label-free quantification using "Quantify Label free (MS2)" option in TPP tools menu. Label-free quantification is discussed further in this chapter under Protein quantification.

## 1.4 Protein quantification and identification

Recent progress on quantitative proteomics technology has enabled researchers to observe quantitative differences in different protein profiles under different experimental conditions and to understand molecular functions of proteins along with their modifications.

### 1.4.1 Label-free quantification

Label-free approach typically carries out detection of peptides, peptide matching across multiple LC-MS data, and selecting biased peptides. Unlike the other quantitation methods, which require stable isotope labeling of proteins, it is not required here and hence effectively calculates relative amounts of proteins in two or more biological samples with greater coverage of peptides. Every protein sample has to be analyzed individually for label-free quantification. This in turn turns out to be relatively fast and cost effective compared with other quantitative methods.

This method of quantification is based on *precursor signal intensity* or *spectral counting*. Precursor signal intensity is measured in case of obtaining high precision mass spectra derived from Orbitrap or time-of-flight (ToF) mass analyzers. The peptide signals are selected at the MS1 level, differentiating quantification from the identification process. Spectral counting, on the other hand, concentrates on counting the spectra identified for a particular peptide derived in various biological samples, and integrates all the quantified peptides belonging to the protein(s).

The end result is in the form of a plot constituting mass to charge and retention time of multiple LC-MS measurements from selected peptide signals. This results in high precision data with peptide signals matched legitimately across runs. There are open-source tools available for label-free quantification like OpenMS, MaxQuant, LFQBench (R-package), and Skyline.

## 1.4.2 Functional annotation and enrichment analysis

This feature is a part of every bioinformatics analysis in omics study. Functional annotation provides biological information on genes related to the proteins, by determining coded and nonprotein-coding genes and elements of these genomes, and further adding biological information to these genomes.

The enrichment analysis employs statistical methods to determine significantly enriched groups of genes. It vouches for genes and proteins overrepresented in a large group of genes that might have affinity with the disease phenotypes. The analysis includes calculating *P*-value that directs the amount to which the proteins set are overrepresented at either the top or bottom of the list, evaluation of statistical significance of a node or pathway based on the *P*-value, and *P*-value for each set is normalized and a false discovery rate is calculated for multiple hypothesis testing. Functional annotation and enrichment analysis of proteins include GO annotation, KEGG annotation, COG annotation, KEGG and Go enrichment, domain enrichment, directed acyclic graph (DAG).

## 1.5 Applications of mass spectrometry

There has been widespread application of mass spectrometry in different areas of science including astrophysics, astronomy, metallurgy, geological, material, and biological sciences. In this chapter we focus on its application in biological and biomedical sciences. The most common application of MSs in biological research is to characterize the macromolecules including proteins, oligonucleotides, oligosaccharides, and lipids. McLafferty and colleagues sequenced up to 100 nucleotides using ESI ion source coupled to FT-ICR MS. DNA methylation can also be analyzed using MS. Isomeric subunits and the branched structure of oligosaccharides make them difficult to analyze compared with proteins and oligonucleotides. Tandem MS is mostly used for oligosaccharides characterization. Lipids namely fatty acid, steroids, and acylglycerol have been identified using MS. The chain length and abundance of lipids present in a biological sample is measured using MS [11].

### 1.5.1 Application in proteomics

1. Determination of molecular mass of peptides and proteins
2. Elucidation of biomolecular structure of proteins
3. Identification of interacting partners of proteins using affinity purified mass spectrometry

### 1.5.2 Application in metabolomics

1. Identification of drug metabolites in blood, saliva, and urine
2. Inborn errors of metabolism in newborn can be screened

### 1.5.3 Application in environment analysis

1. Determination of pesticides in food
2. Monitoring aerosol particles in air
3. Heavy metals leaching
4. Determination of water quality

### 1.5.4 Application in pharmacy

1. Drug discovery
2. Pharmacodynamics and pharmacokinetic analysis
3. Drug metabolism

## 1.5.5 Application in forensic science

1. Analysis of trace evidences like poison or drug
2. Explosive residues identification

## 1.5.6 Applications in medical research

1. Drug testing
2. Neonatal screening
3. Hemoglobin analysis
4. Isotope

## 1.6 Conclusion

Mass spectrometry has revolutionized the field of science, where we can now study science right from stars to the life forms on Earth. Proteins constitute the important part of biological processes and understanding of proteins in detail is crucial for understanding life and its processes. MS has become an essential part of proteomic studies where previously impossible analysis has been made possible with the use of MSs. Currently multiple MSs are available in different configurations and can be assembled as per the requirements. Furthermore, researchers are attracted to develop more advanced and cost effective instruments and also software components for data analysis. MSs generate huge amounts of data and there is need to precisely and accurately derive valuable information from this ocean of data, which can be an added value to the pool of information already available. Hence there is need for biologists to have sound knowledge for analyzing data produced by MSs.

## References

[1] Morris JH, Knudsen GM, Verschueren E, Johnson JR, Cimermancic P, Greninger AL, et al. Affinity purification—mass spectrometry and network analysis to understand protein-protein interactions. Nat Protoc 2014;9(11):2539—54.

[2] Strathmann FG, Hoofnagle AH. Current and future applications of mass spectrometry to the clinical laboratory. Am J Clin Pathol 2011;136:609—16.

[3] Canas B, Lopez-Ferrer D, Ramos-Fernandez A, Camafeita E, Calvo E. Mass spectrometry technologies for proteomics. Brief Funct Genomics 2006;4(4):295—320.

[4] El-Aneed A, Cohen A, Banoub J. Mass spectrometry, review of the basics: electrospray, MALDI, and commonly used mass analyzers. Appl Spectroscopy Rev 2009;44:210—30.

[5] Domon B, Aebersold R. Mass spectrometry and protein analysis. Science 2006;312:212—17.

[6] Zubarev RA, Makarov A. Orbitrap mass spectrometry. Anal Chem 2013;85:5288—96. Available from: https://www.ncbi.nlm.nih.gov/pmc/articles/PMC3017125/.

[7] Baldwin MA. Protein identification by mass spectrometry. Mol Cell Prot 2004;3:1—9.

[8] Michalski A, et al. Ultra high resolution linear ion trap Orbitrap mass spectrometer (Orbitrap Elite) facilitates top down LC MS/MS and versatile peptide fragmentation modes. Mol Cell Prot mcp 2011;11 O111.013698.

[9] Helfer AG, et al. Orbitrap technology for comprehensive metabolite-based liquid chromatographic—high resolution-tandem mass spectrometric urine drug screening—Exemplified for cardiovascular drugs. Anal Chim Acta 2015;891:221—33.

[10] Deutsch EW, Mendoza L, Shteynberg D, Farrah T, Lam H, Tasman N, et al. A guided tour of the trans-proteomic pipeline. Proteomics 2010;10:1150—9.

[11] Finehout EJ, Lee KH. An introduction to mass spectrometry applications in biological research. Biochem Mol Biol Educat 2004;32 (2):93—100.

## Further Reading

Gonzalez-Galarza FF, et al. A critical appraisal of techniques, software packages, and standards for quantitative proteomic analysis. Omics: J Integr Biol 2012;16(9):431—42.

Shteynberg D, et al. iProphet: multi-level integrative analysis of shotgun proteomic data improves peptide and protein identification rates and error estimates. Mol Cell Prot 2011;10(12) M111. 007690.

Siuzdak G. An introduction to mass spectrometry ionization: an excerpt from the expanding role of mass spectrometry in biotechnology. MCC Press: San Diego, JALA: J Associat Lab Automat 2004;9(2):50—63.

# Chapter | 2 |

# Circular dichroism

*Bhaswati Banerjee[1], Gauri Misra[2] and Mohd Tashfeen Ashraf[1]*

[1]*School of Biotechnology, Gautam Buddha University, Noida, India,* [2]*Amity Institute of Biotechnology, Amity University, Noida, India*

## 2.1 Introduction

The determination of secondary and tertiary structure of biomolecules, especially proteins and nucleic acids, is integral to understanding how biomolecules acquire their native, biologically active conformation. Various techniques have been developed that enable researchers to understand how secondary and tertiary structure is formed and which noncovalent interactions are crucial in imparting structural stability. X-ray crystallography is undoubtedly one of the most sensitive techniques that elucidate the structure at atomic level; however the prerequisite of obtaining the biomolecules with high purity required for obtaining the diffraction quality crystals of biomolecules has its own limitations, which allows for the scope of other techniques to be used for studying structural details. Circular dichroism (CD) is one such technique that is routinely employed to understand the secondary as well as tertiary structure of proteins, nucleic acids, and higher structures formed by association of these molecules with their respective ligands. The ease with which a biomolecule's structure and changes therein can be studied with CD has made it a technique of choice for researchers for solution-based studies. As with most other spectroscopic techniques, CD is also primarily used to study, in comparison to native conformation, the structural changes that accompany association/dissociation of ligands and also during unfolding/refolding of the biomolecules. Similar to other spectroscopic methods, the theoretical framework for CD is not fully developed, so the elucidation of raw data is mainly done with the help of certain empirical rules that have been devised based on the study of certain model compounds. Though there are excellent reviews on CD, its applications [1–3], and online databases/tools/repositories [4–11] available, in the following sections an effort has been made to put in simple terms how this technique could be used by researchers in industry and in academics to study structural changes in proteins and nucleic acids.

## 2.2 Principle

The CD essentially deciphers the interaction of plane polarized light (PPL) with an asymmetric molecule. Unlike unpolarized light (whose E vector oscillates in all the planes), the PPL has its E vector oscillating in a single plane, which is achieved by passing unpolarized light through a polarizing material like Polaroid, nicol prism, etc. [12]. From two PPL waves of the same wavelength and amplitude that differs in phase by ¼ of wavelength and whose E vectors are perpendicular to each other, the resultant wave's E vector appears to oscillate in a circular fashion. This is referred to as a right circularly polarized light (CPL), if the oscillation appears to be clockwise to an observer looking at the light source while it is referred to as left CPL if the tip of resultant E vector follows anticlockwise path. If right and left CPLs of equal amplitude are superimposed the result is PPL, while if the two CPL waves are of unequal amplitude the result is elliptically polarized light.

When a symmetric molecule absorbs PPL, both the right and left components are equally absorbed and the emergent light is also a PPL; however, if asymmetric/chiral molecule interacts with PPL, the right and left components of PPL will be unequally absorbed and the resultant wave will be elliptically polarized with CD, that is, $\Delta\mathcal{E} = \mathcal{E}_L - \mathcal{E}_R$. Thus, a CD signal may have both negative and positive values depending on the relative absorption of right ($\mathcal{E}_R$) and left ($\mathcal{E}_L$) components. A word of caution about operation of CD instruments: the nitrogen gas should be flushed through the instrument so as to cool down the machine-lamps; this is very important when the temperature-induced transitions are studied. During such transitions, water is also continuously flushed through the instrument. The nitrogen prevents the formation of ozone caused by the lamp, which can damage the components of the CD. Fig. 2.1 gives an outline of what happens when a PPL wave interacts with a symmetric or asymmetric molecule.

As mentioned earlier, certain empirical rules are generally used to interpret CD data. Following are some of the empirical rules [13]:

1. A CD spectrum is additive; that is, it is the simple sum of the spectra of its components. This is not always strictly true but is certainly a good approximation.
2. The rotational strength (the area under the $\Delta\mathcal{E}$ vs $\lambda$ curve) of a CD curve is a measure of the degree of asymmetry. An agent that increases or decreases these parameters usually does so by increasing or decreasing asymmetry. Although other spectral features usually accompany the change in asymmetry. For example, a commonly used chemical agent, trifluoroethanol

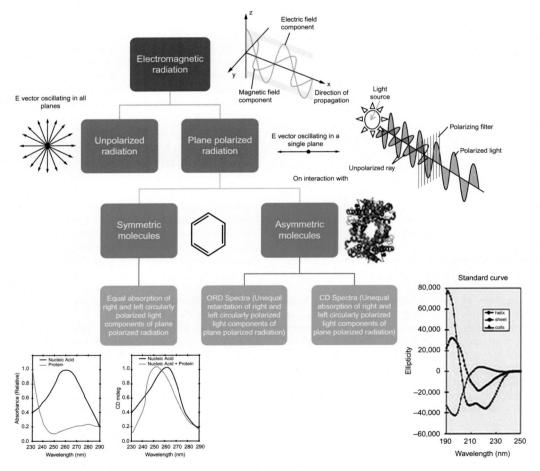

**Figure 2.1** Outcome of interaction of PPL wave with a symmetric/asymmetric molecule.

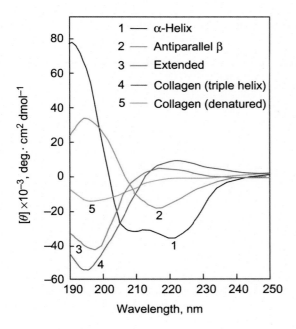

**Figure 2.2** Circular dichroism (CD) spectra of polypeptides and proteins with some representative secondary structures- a, CD spectra of poly-L-lysine in the 1, α-helical (*black*) and 2 antiparallel β-sheet (*red*) conformations at pH 11.1, and 3 extended conformation at pH 5.7, 3 (*green*) [16] and placental collagen in its 4, native triple-helical (*blue*) and 5, denatured (*cyan*) forms [17]. Note that the extended conformation of poly-L-lysine was originally described as a "random coil" but its spectrum is similar to the conformation of poly-L-proline II [18,19], which forms an extended left-handed helix.

(TFE), induces helicity in aqueous proteins; this is accompanied by an increase (more negative) in the intensity of CD signal in the far UV range, that is, 190−250 nm.

3. A chromophore that is symmetric can become asymmetric (optically active) when it is in an asymmetric environment (e.g., a helix). This may or may not be achieved by a change in $\lambda_0$ (wavelength of the peak of the CD curve).

4. The value of $\lambda_0$, (wavelength of maximum signal) and sign of $\Delta\varepsilon$ at $\lambda_0$ allows the chromophore to be identified because it is always very near the value of $\lambda_0$ obtained from simple absorption spectroscopy.

Since the CD signal deals with differential absorption of right and left CPL components of PPL, it is pertinent to mention the range of wavelength used depends on the absorption range of the chromophore. For instance, in

proteins the peptide bond acts as a strong chromophore in the region 180−230 nm (far UV wavelength range) while the aromatic amino acids absorb in the range 250−300 nm (near UV wavelength range).

Also, since the CD deals with differential absorption of right and left CPL components of PPL, the range in which CD signal is obtained coincides with the range of absorption. Fig. 2.2 gives CD spectra of various proteins under different conditions; the details of these are described in the legend for this figure [14,15]. Besides, Table 2.1 above gives an excellent troubleshooting guide for some of the common problems encountered while obtaining the CD data [14].

## 2.3   Raw data analyses

Before we discuss examples to understand the analyses of raw data, certain precautions and important points need to be remembered:

1. The baseline subtraction must be performed by subtracting the CD spectra of the buffer in which our molecule of interest is dissolved from the spectra obtained for compound of interest.

2. Avoid using $H_2O$ that had been stored in a polyethylene bottle for a long time. The polymer additives may elute resulting in the water losing its transparency.

3. Though the protein concentration for observing CD spectra is quite low, around 0.5 mg mL$^{-1}$, it is advisable to remove any chances of unforeseen aggregation by dynamic light scattering of at least by taking the absorbance spectra in wavelength range in which the protein does not absorb, that is, around 350−400 nm.

4. The difference in left and right handed absorbance $A$ ($l$)−$A$($r$) is very small (usually in the range of 0.0001) corresponding to an ellipticity of a few 1/100ths of a degree.

5. Due to high interference with solvent absorption in the UV region, only very dilute, nonabsorbing buffers are used for measurements below 200 nm.

### 2.3.1   Case study 1: to study acid-induced transitions in a protein

Fig. 2.3A shows the far UV CD spectra of cytochrome c, a small protein of about 12,400 Daltons that contains heme prosthetic group [20]. The curve labeled 4 in Fig. 2.3A is far UV CD spectra of cytochrome c in native state while 1, 2, and 3 are acid unfolded state at pH 2.0, in the presence

**Table 2.1 Troubleshooting guide for common problems encountered in CD and their solutions.**

| S. No. | Problem | Solution |
|---|---|---|
| 1. | The CD spectra have very low ellipticity. | Check the protein concentration. |
| 2. | The signal to noise is very low. | Try opening the slit width to 2 nm, as the exact wavelength is not critical. Try increasing the time of data collection. Signal to noise increases as the square root of the time of the signal averaging time. |
| 3. | The samples precipitate when heated. | Try using a different buffer or include stabilizing agents such as low concentrations of glycerol. |
| 4. | The samples refold to give the same spectra as the starting material but the unfolding and refolding curves are displaced from each other. | The samples are not at thermal equilibrium during the measurements. Try increasing the equilibration time. |
| 5. | The curve of ellipticity as a function of temperature appears to have a typical sigmoidal shape, but none of the folding equations fit the data and the error of the fit is large. | Make sure that the initial parameters are close to the actual values of the data and that the correct units are used for the initial estimates of the ellipticity of the folded and unfolded protein (e.g., millidegrees or mean residue ellipticity). Make sure the initial estimate of the $T_M$ of folding is close to the midpoint of the transition. If this doesn't work, try increasing or decreasing the initial estimates of the enthalpy of folding. |
| 6. | The calculated $T_M$ values for different concentrations of protein are not close to each other, even when the data are modeled to fit the dissociation of dimers or trimers. | Determine the oligomerization state of the protein using independent methods such as gel filtration or ultracentrifugation, and use the van't Hoff equation to determine the thermodynamic parameters. |
| 7. | The macro doesn't work properly. | Check the users manual to make sure you have programmed the machine properly. Usually machines will have sample macros. Try to program a single temperature step with using only one or two samples. |
| 8. | A. When the spectra are deconvoluted using the CCA algorithm some basis curves only contribute to a spectrum obtained at a single temperature.<br>B. When the spectra are analyzed using the CCA algorithm, some of the curves have very similar shapes but are displaced from each other.<br>C. When the data is deconvoluted using singular value decomposition (SVD), while it is clear there are more than two states, only one principal component is resolved. | A. If a spectrum is noisy and has outlying points, the CCA algorithm will identify the spectrum as a unique basis set. Try removing the spectrum from the data set.<br>B. The CD spectra of the cells may change as a function of temperature, leading to shifts in the baseline. Obtain spectra of the cuvettes filled with water as a function of temperature and correct the individual data set for the contribution of the cells.<br>C. If the spectra of the folded and unfolded proteins have maxima (nodes) that are identical to each other, SVD the solution may be "singular" and only one spectrum will deconvoluted. |

Source: Adapted from Greenfield NJ. Using circular dichroism collected as a function of temperature to determine the thermodynamics of protein unfolding and binding interactions. Nat Prot 2006;1(6):2527–35.

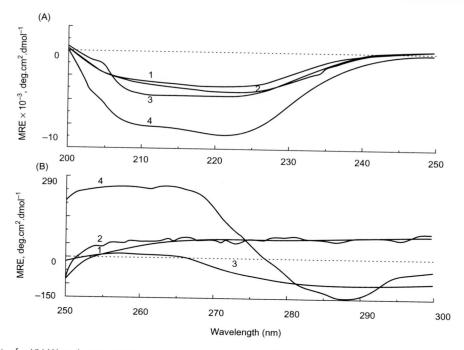

**Figure 2.3** Cyt c far-UV (A) and near UV CD spectra (B): (1) acid unfolded state at pH 2.0, 10 mM glycine HCl buffer; (2) 33 mM TCA-induced state; (3) 3.3 mM TCA-induced state; (4) native state at pH 7.0, 10 mM sodium phosphate buffer. Protein concentration was 26 (A) and 50 μM (B). The path length was 0.1 (A) and 1 cm (B).

of 33 mM TCA and in the presence of 3.3 mM TCA, respectively. Since cytochrome c is an α-rich protein, two shoulders are observed in far UV CD spectra, at 208 and 222 nm, which is a characteristic feature of proteins that have a significant amount of α-rich regions. It can be seen that the intensity of signal in all three unfolded states is far less as compared with the native state. Note that the far UV CD spectrum gives details about the secondary structure of biomolecules (i.e., primarily stabilized by hydrogen-bonding interactions between amino acids that lie close together in primary sequence). As mentioned above, the main chromophore in this region is the amide linkage of peptide bond. On the other hand, the near UV CD spectrum gives a rough idea of the overall three-dimensional structure of proteins; the main chromophores in this region being the aromatic amino acids, that is, tryptophan, tyrosine, and to some extent phenylalanine. Fig. 2.3B above shows the near UV CD spectrum of cytochrome c under conditions similar to Fig. 2.3A. As near UV CD spectra gives the global picture, it is difficult to deduce the contribution/influence of specific aromatic amino acid. Therefore, the near UV CD spectra are generally used to observe the change in the spectrum between

different experimental conditions. Far UV CD spectra, in contrast, may be used for more quantitative information, in terms of percentage of residues involved in forming the helix. The α-helical content of a protein can be calculated from the MRE value at 222 nm using the following equation as described by Chen et al. [21]:

$$\% \, \alpha - \text{helix} = \frac{\text{MRE}_{222} - 2340}{30,300 \times 100}$$

where MRE is mean residue ellipiticity; $\text{MRE}_{222}$ is mean residue ellipticity at 222 nm.

### 2.3.2 Case study 2: to study pH-induced transitions in a protein

In one of the studies conducted to study the effect of alkaline pH on Concanavalin A (Con A) in the presence/absence of metal ion $(Mg^{2+})$ [20], certain interesting results were obtained as shown in Fig. 2.4. It shows the far-UV CD spectra of Con A under different pH conditions between pH 7 and 12. The spectrum of Con A at pH 12 shows the structural transition with a CD band at 212 nm (Curve 5), a shift

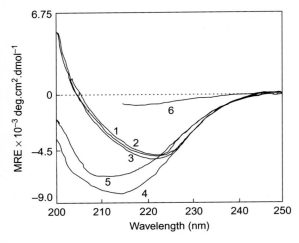

Figure 2.4 Far-UV CD spectra of Con A at pH 7 (Curve 1); pH 7, metallized (Curve 2); pH 7 demetallized (Curve 3); pH 12 (Curve 5); pH 12, demetallized (Curve 4) and in the presence of 6 M Gdn HCl (Curve 6).

of around 11 nm from that of 223 nm for native Con A at pH 7 (Curve 1). The signal at 217 nm at pH 12 was also considerably enhanced as compared with that at pH 7. Though the presence of metal ion/EGTA at pH 7 was not found to significantly affect the signal in the far-UV region (Curve 1 and 2 respectively) but the presence of metal ions at pH 12 was found to aggregate the protein. On the other hand, the presence of EGTA at pH 12 (Curve 4) showed similar transition as that of apoprotein at pH 12 (Curve 5) but with enhanced signal intensity. Curve 6 shows the far UV CD spectra of Con A denatured in the presence of 6 M Guanidine hydrochloride (Gdn HCl).

### 2.3.3 Case study 3: to study structural transitions in nucleic acids

The structural changes in nucleic acids, DNA and RNA, can also be studied using CD spectroscopy. Though the nitrogenous bases per se are not chiral but the chirality gets induced due to nearby chiral sugar present in ribo/deoxyribonucleotides. This induction of chirality is illustrated in Fig. 2.5 below.

In DNA, the helical structure can be experimentally determined by CD spectra. The sign and shape of the CD spectra are different for B-DNA, which has a right-handed double-helical structure as compared with the Z-DNA, which has left-handed double-helical structure. In

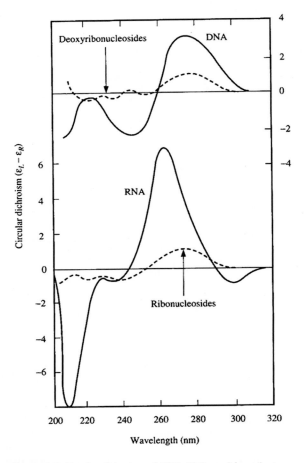

Figure 2.5 Circular dichroism of DNA, RNA, and free ribo/deoxyribonucleotides.

particular, for B-DNA the signal at around 295 nm in CD spectra is positive while in the case of Z-DNA it is just the opposite. Following are three important factors that affect the CD spectra of DNA: the conformation of dG monomer, the hydrogen-bonding interaction between two helices, and lastly, the stacking interactions between nucleic acid bases.

### 2.3.4 Case study 4: determination of $T_m$ and other thermodynamic parameters

CD is one of the simplest spectroscopic techniques used for measuring the thermal stability of proteins in natively folded and different partially folded/unfolded conformations. Using CD, the fraction of molecule that is in native/denatured state under given conditions can be calculated. For this, the MRE values at particular wavelength (e.g., at 222 nm for an alpha-rich protein) are used to calculate fraction denatured ($f_D$), which is then plotted against temperature/pH/denaturant concentrations. Assuming two-state protein folding, denaturation midpoint is defined as that temperature ($T_m$) or denaturant concentration ($C_m$) at which both the folded and unfolded states are equally populated at equilibrium. $T_m$ is often determined using a thermal shift assay. For calculating melting temperature of a protein, let's consider a protein and a single point mutation in two states, native (folded) and denatured (unfolded) [22]. The equilibrium is defined like any other reaction: $K_{eq}$ = [denatured protein]/[native protein]. If one uses PPL or fluorescence to determine the fraction of a protein's folded or unfolded conformations at different temperatures, a protein-melting curve can be produced [23]. To determine the folded fraction from the thermal denaturation experiments, the following formula can be used [23]:

$$\frac{[\theta]^{obs} - [\theta]^{den}}{[\theta]^{nat} - [\theta]^{den}}$$

where $[\theta]^{obs}$ is the ellipticity at a given temperature, $[\theta]^{den}$ represents the ellipiticity at highest temperature and $[\theta]^{nat}$ at lowest temperature, respectively. An extensively studied example for the usage of CD in understanding the thermal stability profile of a chaperone is Pf Hsp70-1 from *Plasmodium falciparum* and its truncated variants. The main transition is observed in the range of 30−45°C, thereafter loss of intensity at $\theta_{222}$ nm followed by a second transition at 80°C (Fig. 2.6A). A single change in thermal denaturation curve was observed with $T_m$ value of 45°C for its nucleotide binding domain (NBD), pointing towards cooperative unfolding with no intermediate stages and following the standard two-state model of unfolding (Fig. 2.6B). Its other domain, namely substrate binding domain (SBD), shows transition at approximately 40°C, retaining 80% of the compact structure (Fig. 2.6C). SBD combined with the C-terminal domain of this protein exhibits folded structure till 80°C, which later causes a decline in $\theta_{222}$ nm intensity (Fig. 2.6D). These results have remarkably established that the structural stability of the PfHsp70-1 is contributed majorly by the C-terminal domain in complex with the SBD as the latter truncated mutants enhanced the stability of the otherwise unstable NBD, reflected from CD studies (Fig. 2.6E).

CD is also used to study the structural changes induced in Pf Hsp70-1 on chemical denaturation using urea

(Fig. 2.7) and GdnCl (Fig. 2.8) by monitoring the changes in the ellipticity at $\theta_{222}$ nm in the presence of increasing concentrations of respective denaturants [23]. The above examples clearly demonstrate the application of CD for the determination of structural changes and stability in response to the changes in the immediate environment. Such applications are extensively useful for gaining structural and functional insights of proteins.

Both the enthalpy ($H$) and entropy ($S$) can be calculated using a curve of a fraction of protein in folded and unfolded state. The S and H values can be calculated using the van't Hoff equation and plot. The van't Hoff equation explains the temperature dependence of the equilibrium constant. To use this equation, the protein must refold—this is an equilibrium problem! The van't Hoff equation is derived from Gibb's free energy equation: $\Delta G = \Delta H - T\,\Delta S$

With the understanding that for this reaction of folding and unfolding: $G = -RT\,\ln K_{eq}$, one can create the van't Hoff equation to relate equilibrium constant to temperature by substituting the two equations and rearranging them to generate the van't Hoff equation:

$$\ln K_{eq} = \frac{-H}{R\left(\frac{1}{T}\right) + \left(\frac{S}{R}\right)}$$

Since the equation is a straight line equation, the plot of $\ln K_{eq}$ versus $1/T$, known as the van't Hoff plot, yields a straight line with slope $-H/RT$ and intercept on Y-axis = $S/R$. Using both the melting or transition curve and van't Hoff's plot and equation, one can determine the thermodynamic functions of protein stability (the fraction of protein at a given temperature that is native or denatured). The van't Hoff plot helps determine the fraction of protein folded from the transition curve where [native protein] = [denatured protein] and convert that information to $K_{eq}$ for each temperature. From this data one creates a van't Hoff plot and calculates the enthalpy and entropy from the slope and Y intercept.

To determine the stability of a protein accompanying a change (ligand binding, protein interaction or mutation), one needs to determine the $\Delta G$ for each conformational transition, for instance in the case of development of mutant, $\Delta G$ (protein) = $G$ wild-type−$G$ mutant. A positive value gives clear indication that the unfolding of wild type protein is less favorable than the mutant as per the calculated value, thereby meaning that the mutation decreases the stability of the native protein. On the other hand, a negative value indicates the unfolding of the wild type is more favorable than the mutant or that the mutation stabilizes the native structure.

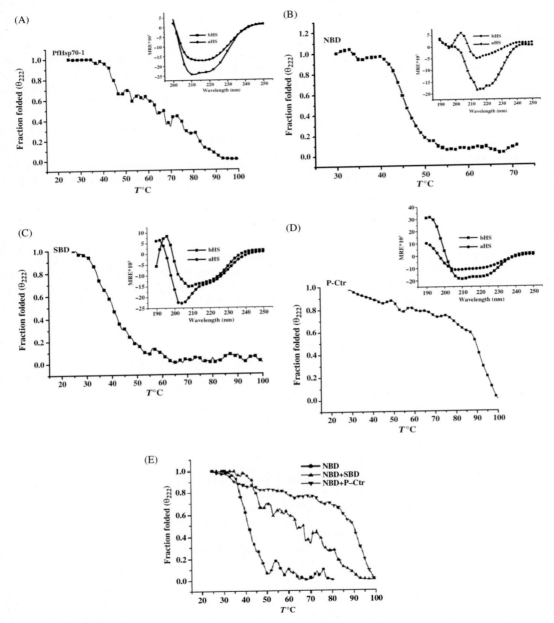

**Figure 2.6** Circular dichroism-based thermal stability analysis of native and truncated mutants of (A) PfHsp70-1, (B) NBD, (C) SBD, and (D) SBD along with C-terminal tail. Inset depicts far-UV spectra for each of the constructs before (bHS) and after (aHS) heat denaturation. (E) Folded fraction of NBD, SBD + NBD and SBD with C-terminal + NBD in equimolar ratio as measured by changes in CD ellipticity at 222 nm.
*Adapted from Misra G, Ramachandran R. Hsp70-1 from Plasmodium falciparum: Protein stability, domain analysis and chaperone activity. Biophys Chem 2009;142:55−64.*

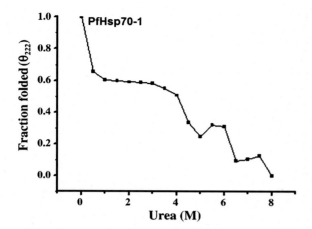

**Figure 2.7** Changes in PfHsp70-1 $\theta_{222}$ with increase in urea concentration.
*Adapted from Misra G, Ramachandran R. Hsp70-1 from Plasmodium falciparum: Protein stability, domain analysis and chaperone activity. Biophys Chem 2009;142:55–64.*

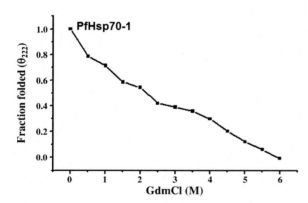

**Figure 2.8** Changes in PfHsp70-1 $\theta_{222}$ with increase in GdnCl concentration.
*Adapted from Misra G, Ramachandran R. Hsp70-1 from Plasmodium falciparum: Protein stability, domain analysis and chaperone activity. Biophys Chem 2009;142:55–64.*

## 2.4  Miscellaneous examples

Since the β-rich aqueous proteins have their characteristic far UV CD spectrum, the changes in their structure can be easily studied by CD. For instance, Con A is an extensively studied lectin that is rich in β-rich regions; around

64% of its residues are engaged in forming β-pleated sheets. The native conformation of Con A is a tetramer that is composed of a dimer of two dimers. The tetramer has a characteristic trough around 223 nm in far UV CD region; any substantial change in secondary structure of Con A is accompanied by a change in CD signal at 223 nm. Thus, the changes in secondary structure of both α- and β-rich proteins can be studied by far UV CD spectroscopy.

The effect of certain helix-inducing agents like TFE and HFIP (hexefluoroisopropanol) can be studied using the far UV CD spectroscopy. Apart from solvent perturbation studies, the change in structure on binding of ligand (e.g., drug molecule, like an enzyme inhibitor) to a protein can be studied by CD spectroscopy.

## 2.5  Conclusion

CD is an important noninvasive spectroscopic technique for understanding the structure of biomolecules in solution and therefore plays an important role in establishing the structure–function relationship in biomolecules like proteins. Relative changes in structure due to influence of environment on sample (pH, denaturants, temperature, etc.) can be monitored very accurately. In particular, the fine details of secondary structure as well as changes in the tertiary structure of biomolecules like proteins can be studied. For α-helical rich proteins the number of residues involved in helix formation can be determined. Though the application of near UV CD spectroscopy has not been that instrumental, it is being successfully used to monitor the changes in the overall three-dimensional structure of protein.

## Acknowledgments

The authors have no conflict of interest. The authors are thankful to Priyansh Srivastava, Fauzan Ahmed from Amity Institute of Biotechnology, Amity University, Noida and Srishti Jha, School of Biotechnology, Gautam Buddha University for their valuable contribution in preparation of this work, especially the illustrations, which convey points that would otherwise be difficult to put forth in words. The authors are also thankful to Gautam Buddha University for providing the required infrastructure.

# References

[1] Sreerama N, Woody RW. Computation and analysis of protein circular dichroism spectra. Methods Enzymol 2004;383:318—51.

[2] Gillard RD. Circular dichroism- a review. Analyst 1963;88:825—8.

[3] Whitmore L, Wallace BA. Protein secondary structure analyses from circular dichroism spectroscopy: methods and reference databases. Biopolymers 2007;89(5):392—400.

[4] Miles AJ, Wallace BA. CDtoolX, a downloadable software package for processing and analyses of circular dichroism spectroscopic data. Prot Sci 2018. Available from: https://doi.org/10.1002/pro.3474.

[5] http://cddemo.szialab.org/. An excellent online tutorial on electromagnetic waves and circular dichroism.

[6] Wiedemann C, Bellstedt P, Görlach M. CAPITO - a web server based analysis and plotting tool for circular dichroism data. Bioinformatics 2013;29(14):1750—7.

[7] Whitmore L, Wallace BA. DICHROWEB, an online server for protein secondary structure analyses from circular dichroism spectroscopic data. Nucl Acids Res 2004;32 (Web Server issue). Available from: https://doi.org/10.1093/nar/gkh371.

[8] CD Pro- A Software Package for Analyzing Protein CD Spectra (2004). https://sites.bmb.colostate.edu/sreeram/CDPro/

[9] Whitmore L, Woollett B, Miles AJ, Klose DP, Janes RW, Wallace BA. PCDDB: the protein circular dichroism data bank, a repository for circular dichroism spectral and metadata. Nucl Acids Res 2011;39 (Database issue):D480—6.

[10] Sreerema N, Woody RW. A self-consistent method for the analysis of protein secondary structure from circular dichroism. Anal Biochem 1993;209:32—44.

[11] Sreerama N, Woody RW. Estimation of protein secondary structure from circular dichroism spectra: comparison of CONTIN, SELCON, and CDSSTR methods with an expanded reference set. Anal Biochem 2000;287:252—60.

[12] Wilson K, Walker J, editors. Principles and techniques of biochemistry and molecular biology. Publisher Cambridge University Press; 2010.

[13] Freifelder D. Physical biochemistry: applications to biochemistry and molecular biology. Publisher W. H. Freeman; 1982.

[14] Greenfield NJ. Using circular dichroism collected as a function of temperature to determine the thermodynamics of protein unfolding and binding interactions. Nat Prot 2006;1(6):2527—35.

[15] Greenfield NJ. Using circular dichroism spectra to estimate protein secondary structure. Nat Prot 2006;1:2876—90.

[16] Greenfield NJ, Fasman GD. Computed circular dichroism spectra for the evaluation of protein conformation. Biochemistry 1969;8:4108—16.

[17] Bentz H, Bächinger HP, Glanville R, Kühn K. Physical evidence for the assembly of A and B chains of human placental collagen in a single triple helix. Eur J Biochem 1978;92:563—7.

[18] Tiffany ML, Krimm S. Effect of temperature on the circular dichroism spectra of polypeptides in the extended state. Biopolymers 1972;11:2309—16.

[19] Woody RW. Circular dichroism and conformation of unordered poly-peptides. Adv Biophys Chem 1992;2:37—79.

[20] Ashraf MT. Spectroscopic characterisation of folding intermediates of various proteins and purification of lectin from Trigonella foenmgraecum: a thesis in Biotechnology. Aligarh Muslim University, Aligarh, 2008.

[21] Chen YH, Yang JT, Martinez HM. Biochemistry 1972;11:4120.

[22] http://home.sandiego.edu/~josephprovost/VantHoffplots.pdf

[23] Misra G, Ramachandran R. Hsp70-1 from Plasmodium falciparum: protein stability, domain analysis and chaperone activity. Biophys Chem 2009;142(2009):55—64.

# Chapter | 3 |

# Fluorescence spectroscopy

*Gauri Misra*
*Amity Institute of Biotechnology, Amity University, Noida, India*

## 3.1 Introduction

Protein structures are mainly the combined clusters of 20 amino acid permutations in a three-dimensional space. Correlation between the protein three-dimensional structures and their functions holds an essential importance in the field of computational drug discovery. Therefore, the knowledge of native protein structures is of utmost importance. The precise practices like X-ray crystallography and electron microscopy often do not successfully capture the dynamics of protein structure—function relationship. Techniques like fluorescence spectroscopy are less invasive in terms of preserving the native structure of the proteins while studying their three-dimensional conformations.

Fluorescence spectroscopy exploits the phenomenon of electron excitation upon collision with high energy particles like photons and other excited electrons and emission of photons while lowering their energy down to the ground state. The molecules showing the fluorescence activity are termed as fluorophores. In biophysical studies of the macromolecules such as proteins and nucleic acids these fluorophores act as a physical marker for their structural studies. The fluorophores could be either extrinsic like radioactive probes and dyes or intrinsic like specific amino acids in protein chains. The intrinsic fluorophores are less expensive and do not require any foreign intervention like in the case of extrinsic fluorophores.

## 3.2 Principle

Fluorescence refers to the emission of light on electronic transition from singlet excited state to singlet ground state.

The incident light rays on molecules results in the absorption of energy that causes conformational changes, leading to vibrational relaxation (the lowest vibrational level). If the aromatic molecule is rigid and cannot relax vibrationally to the ground state, then it reaches ground state by emission of light. This emission of light causes fluorescence. The incident light is first absorbed by the $\pi$ electron systems of these molecules. The system's electrons are excited from the ground state (S0) to the excited energy state (S1) (Fig. 3.1). Additionally these electrons jump to different vibrational levels of the excited state. The electrons via thermal consumption of the energy also go to the lowest vibrational energy level of the excited state. Since these molecules are capable of fluorescing, the electrons jump from the lowest energy level of the excited state to different vibrational energy levels of the ground state through the emission of specific quantums of energy in the form of light [1].

## 3.2.1 Instrumentation

An external light source provides a photon of energy that is absorbed by a fluorescent molecule raising it to an excited state. The excited state exists for a finite time before relaxing back to the ground state resulting in the release of a photon that generates fluorescence emission [2]. A molecule in its electronic and vibrational ground state can absorb photons matching the energy difference of its various discrete states. The required photon energy has to be higher than that required to reach the vibrational ground state of the first electronic excited state. The excess energy is absorbed as vibrational energy and quickly dissipates as heat by collision with solvent molecules. The molecule thus returns to the vibrational ground

**Data Processing Handbook for Complex Biological Data Sources.** DOI: https://doi.org/10.1016/B978-0-12-816548-5.00003-4

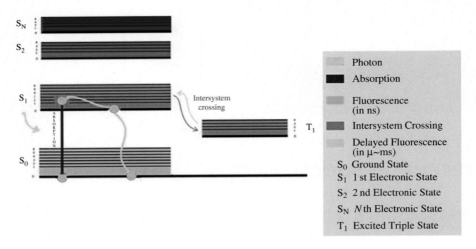

**Figure 3.1** Jablonski diagram illustrating the three-stage fluorescence process.

state. These nonradiating transitions from one energetic state to another with lower energy are the relaxation processes, termed as internal conversion (IC). The ground state (S0) is achieved by the molecule transitioning from the lowest level of the first electronic excited state by either emitting fluorescence light or by a nonradiative transition. Upon radiative transition, the molecule can end up in any of the vibrational states of the electronic ground state (as per quantum mechanical rules). If there exists an overlapping of the ground state vibrational levels with those of the electronic excited state, there will be no fluorescence emission. It will lead the molecule to return to the ground state by nonradiative IC causing the dissipation of the excitation energy. Fluorophores refers to the presence of fluorescent group in a molecule [3]. Besides the fluorescence phenomenon, there is intersystem crossing (ISC) when the molecule transitions to the triplet state from an electronically excited singlet state, driven by spin−orbit coupling between orbits of variable multiplicities. It is a nonradiative change between states of different multiplicity. ISC involves spin inversion of the electrons in the excited causing the two unpaired electrons with same orientation, spin value of 1 and triplet state. The transitions between states of different multiplicity are not permissible. Emission rates of fluorescence correspond to $10^8 \, s^{-1}$ with a fluorescence lifetime of about 10 ns $(10 \times 10^{-9} \, s)$. An important concept of Stokes shift is applicable in fluorescence where a molecule on absorption of photon having higher frequency/energy or shorter wavelength emits the photon exhibiting lower energy/frequency or longer wavelength. The vibrational relaxation

and reorganization of the solvent molecules are the major factors contributing to Stokes shift [4].

The fluorescence spectrophotometer consists of a light source, two monochromators (one helps in the choice of the excitation wavelength and the other is used for analyzing the emitted light), a sample holder and a detector placed perpendicular to the excitation beam. When the sample molecules undergoes excitation, it causes the emission of fluorescence in all possible directions, which is detected by the photocell present at 90 degrees to the excitation beam. Xenon arc lamps are one of the common light sources emitting radiation in the UV, visible, as well as the near-infrared regions. This light from the source is directed using an optical system to the excitation monochromator that permits either preselection of a defined wavelength or scanning of certain wavelength range. The exciting light then passes through the sample chamber containing fluorescence cuvette of quartz or glass. The excited light beam passing through the sample cell causes the excitation of molecules present in the solution. This also causes a fraction of light emitted by the excited molecules. Light emitted at right angles to the incoming beam is analyzed by the emission monochromator. Intensity of fluorescence at preselected wavelength helps in the wavelength analysis of emitted light. The analyzer monochromator directs the emitted light of the preselected wavelength to the detector which can be a photomultiplier tube that measures the light intensity. The output current from this photomultiplier is directed towards a measuring device that indicates the extent of fluorescence (Fig. 3.2) [1,5].

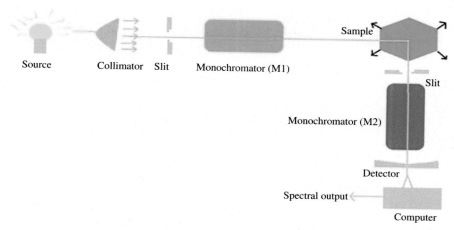

**Figure 3.2** Diagrammatic representation of the instrumentation.

## 3.3 Intrinsic fluorescence

Intrinsic fluorophores in proteins are constituted of individual amino acids. The list includes tryptophan, tyrosine, and phenylalanine respectively. The tryptophan and tyrosine are used in biophysical studies of protein structure more as compared with the phenylalanine, as phenylalanine produces less fluorescence in comparison to the other two aromatic amino acids. The tyrosine and tryptophan can be excited by irradiation with the wave having wavelength equivalent to the 280 nm. Moreover, the tryptophan can also be excited by wavelength equivalent to 295 nm. For the analysis of the data, the plot between fluorescence intensity versus wavelength and fluorescence versus concentration is studied. Changes in the fluorescence intensity are observed based on the environment of Trp and Tyr residues in the protein. A high quantum yield in terms of fluorescence intensity is observed when these residues are buried in a hydrophobic environment (generally in the folded state) and vice versa in a hydrophilic surroundings when these residues are exposed to the solvent (in unfolded state of protein).

Although nucleic acids have significantly weaker absorption in the visible versus UV spectrum, they exhibit low, but detectable, absorption due to the electron delocalization effect, in part arising from the aromatic ring, that is, purines. There is a significant enhancement in fluorescence emission signal in the visible range when the concentration of nucleic acid reaches the level of that present in the interphase nuclei and metaphase chromosomes [6].

## 3.3.1 Protein stability studies

The fluorescence studies can be effectively utilized for the determination of protein stability. For instance, deciphering the role of the C-terminal domain of *Plasmodium falciparum* Hsp 70-1 (PfHsp70-1) protein in the cytoprotection of the malarial parasite was delineated using fluorescence based denaturation studies. Intrinsic tryptophan based studies were used to observe the changes in the tertiary structure of this protein, which harbors three tryptophan residues at 32, 101, and 593. The truncated mutants of the proteins that were constructed in such a manner that the first two tryptophan residues were confined to the N-terminal ATPase domain (NBD), whereas the third Trp residue was the part of long C-terminal domain (P-Ctr). The central substrate binding domain (SBD) was devoid of any tryptophan residues. A range of denaturant concentrations that included 0−8 M urea and 0−6 M guanidium chloride was used to incubate the proteins for different time intervals spanning from 6 to 8 hours so as to reach the equilibrium. The samples were excited at a wavelength of 290 nm and the emission spectrum was recorded from 300 to 400 nm. The different protein constructs' emission maximum was located between 335 and 345 nm. It is important to mention that a buried Trp residue is known to give the emission maximum in the range 330−340 nm. However, a Trp residue exposed to the solvent exhibits the emission maxima in a higher wavelength of around 353 nm. These emission spectra of PfHsp70-1 and its various truncated mutants showed a red shift (emission maxima in the direction of longer wavelength) indicating clearly that the tryptophan residues were present inside hydrophobic pockets that were buried with appropriate protein folding. This red shift was gradual in case of both

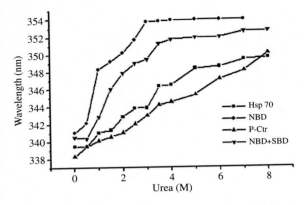

**Figure 3.3** Red shift in Trp fluorescence emission wavelength of PfHsp70-1 and various mutants on incubation with increasing concentrations of urea.
*Adapted from Misra G, Ramachandran R. Hsp70-1 from* Plasmodium falciparum: *protein stability, domain analysis and chaperone activity. Biophys Chem 2009;142(1–3):55–64.*

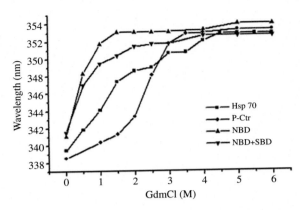

**Figure 3.4** Red shift in Trp fluorescence emission wavelength of PfHsp70-1 and various mutants on incubation with increasing concentrations of GdmCl.
*Adapted from Misra G, Ramachandran R. Hsp70-1 from* Plasmodium falciparum: *protein stability, domain analysis and chaperone activity. Biophys Chem 2009;142(1–3):55–64.*

full length protein and P-Ctr with emission maxima centered at 350 and 351 nm respectively at 7 M concentration of urea (Fig. 3.3). The structural transitions were more noticeable in the full length protein as compared with the truncated mutant P-Ctr on incubating with low urea concentrations. It was observed that the Trp exposure to solvent present in NBD was complete at 3 M urea concentrations but the presence of equimolar SBD leads to solvent exposure of Trp at 5 M urea. This transition in the full length and P-Ctr is achieved at a higher concentration of 7 M urea (Fig. 3.3). This red shift in emission maxima is indicative of a more stable SBD and P-Ctr contributing towards the parasite cytoprotection [7].

The stability studies were also performed in the presence of GdmCl so as to confirm the observations achieved implying urea denaturation studies. The full length protein shows a red shift with emission maxima corresponding to 352 nm and 4.5 M GdmCl concentrations. However, individual NBD lost its secondary structure at 1 M GdmCl. Addition of equimolar concentration of SBD to NBD shifted the denaturation to 2.5 M GdmCl concentrations (Fig. 3.4) [7].

## 3.3.2 Protein interaction studies

The interaction of Pf Hsp70-1 with the peptides endowed with transit peptide-like features was determined using saturation binding experiments. Keeping a fixed concentration of the protein, a range of peptide concentration was incubated with the protein. Both intrinsic and extrinsic

fluorescence studies were used for the determination peptide binding affinities toward the target protein. The protein was subjected to excitation wavelength of 295 nm to exploit intrinsic fluoresce properties of the three Trp residues (32, 101, 593) and an emission spectrum was recorded from 300 to 420 nm. The equilibrium binding constant $K_d$ was determined using the following equation:

$$Y = B_{max} \times X / (K_d + X)$$

where $B_{max}$ represents the maximum specific binding in similar units as $Y$ and $K_d$ is the equilibrium binding constant having same units as $X$.

The emission maxima peak at 341 nm was indicative of the buried Trp. The increasing concentrations of peptides on titrating with a fixed concentration of the protein did not show any change in the emission peak position, however, changes in fluorescence intensities were observed when different peptides interacted with PfHsp70-1 (Fig. 3.5). The saturation isotherms obtained by plotting percent change in fluorescent intensity versus peptide concentration were used for determining the $K_d$ (dissociation constant) values [8].

## 3.4 Extrinsic fluorescence

Proteins that do not contain inherent Trp or Tyr residues can be tagged to external fluorophores such as FITC (fluorescein isothiocyanate), Alexa red, etc. Some proteins bind to cofactors namely NADH and FAD that exhibit

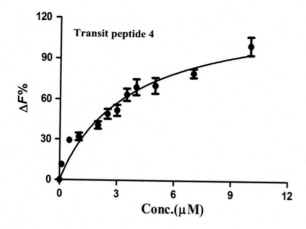

**Figure 3.5** Saturation isotherms for interaction of peptide 4 with PfHsp70-1. The percent change in fluorescence intensity at 341 nm (the emission maxima of PfHsp70-1) plotted as a function of peptide concentration. Error bars represent the SE of the corrected mean.

*Adapted from Misra G, Ramachandran R. Exploring the positional importance of aromatic residues and lysine in the interactions of peptides with the* Plasmodium falciparum *Hsp70-1. Biochim Biophys Acta (BBA)-Prot Proteom. 2010;1804(11):2146–52.*

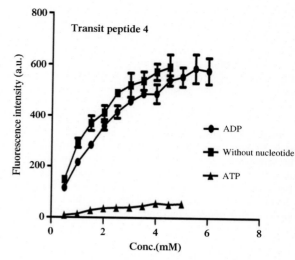

**Figure 3.6** Binding affinity of FITC-labeled peptide 4 with PfHsp70-1 observed from saturation binding in the presence of 1 mM of the respective nucleotides (ADP and ATP) at constant protein concentration. Saturation curves in the absence of the respective nucleotides are also shown. Error bars depict the SE of the corrected mean [8].

**Table 3.1  List of few extrinsic fluorophores.**

| Fluorophores | Absorption wavelength | Emission wavelength | References |
|---|---|---|---|
| Salarg and Mn-Salarg | Shows three prominent peaks at around 215, 256, and 320 nm | 428 nm in methanol and 508 nm in water [6] | [9] |
| Fluorescein isothiocyanate (FITC) | 276 nm | 518 nm [2] | [10] |
| Rhodamine | 530 nm | 510 nm to around 710 nm with the peak at 550 nm [10] | [11] |

fluorescence properties. Compounds like ANS and TNS bind to the hydrophobic areas on proteins that get exposed on protein unfolding. The extrinsic fluorescence studies were effectively used to cross check the binding affinity of PfHsp70-1 with the different peptides. The FITC-labeled peptides were used for extrinsic fluorescence studies. The emission spectra were recorded from 400 to 620 nm with a peak centered at 490 nm. The effect of nucleotides ADP and ATP on the structure of PfHsp70-1 was also studied using extrinsic fluorescence as intrinsic fluorescence could not be recorded because the presence of nucleotides will contribute itself toward quenching of

fluorescence (Fig. 3.6). The $K_d$ values were quite useful in understanding the enhancement of peptide affinity for this chaperone in the presence of ADP [8].

A list of commonly used extrinsic fluorophores used for various purposes are summarized in Table 3.1.

## 3.5  Fluorescence polarization

The time dependent polarization experiments hold considerable relevance in the study of DNA and protein conformations at the single-molecule level. Fluorophores are usually dipolar in nature, therefore their orientation is important as it affects the absorption and emission phenomenon. The excitation rate is governed by the incident power, absorption cross-section $\sigma$, relative orientation of the incident electromagnetic field $E$, and the absorption dipole moment $\mu_{abs}$, given by the following equation:

$$k_e = \sigma(\mu_{abs} \cdot E)^2$$

Intensity of emission is dependent on both the excited state S1 population and the detection efficiency that can be sensitive to polarization. The two components of the linearly polarized intensities namely $I_s$ and $I_p$ are along two perpendicular directions in a plane orthogonal to the direction of light propagation (Fig. 3.7); the polarization anisotropy dependent on time is derived as:

$$r = I_s - I_p I_s + 2I_p$$

On the other hand for a molecule that is moving freely in all directions it is depicted as follows:

$$r(t) = 1/5\ (3\cos2\alpha - 1)e - t/\tau r$$

where $\alpha$ is the angle between the emission and the absorption dipole of the molecule, $\tau r$ is the rotational diffusion time [12].

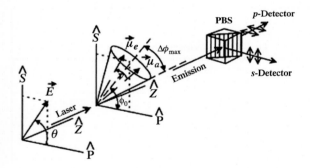

**Figure 3.7** Polarization spectroscopy geometry. $\vec{E}$ is the electric field at an angle $\theta$ with the $p$ polarization axis. The excitation propagates along collection axis $z$, $\vec{\mu}_a$ and $\vec{\mu}_e$ are the absorption and emission dipole moments, initially aligned. $\nu$ represents the rotational diffusion of the emission dipole during the excited lifetime. The dipole is expected to reside in a cone positioned at an angle $\varphi 0$ projected on the $(s, p)$ plane with a half-angle $\Delta\varphi$max. A polarizing beam splitter splits the collected emission in two signals $I_s$ and $I_p$, that are simultaneously recorded by APDs.
*Adapted from Michalet X, Weiss S, Jäger M. Single-molecule fluorescence studies of protein folding and conformational dynamics. Chem Rev 2006;106(5):1785–813.*

## 3.6 Fluorescence resonance energy transfer

In this process a donor fluorophore in the excited electronic state and a neighboring acceptor chromophore are involved in nonradiative energy exchange using dipole–dipole interactions. Both the donor and the acceptor function as a coupled dipole oscillating with a similar resonance frequency. The distance between donor and acceptor is generally a few nanometers. Rate of fluorescence resonance energy transfer (FRET) referred to as kFRET is derived as follows:

$$kFRET = 1\tau 0\ (R0R)\ 6$$

where $\tau 0$ is the donor fluorescence lifetime in the absence of acceptor and $R0$ is the Förster radius (its value ranging from 2 to 6 nm for commonly used dye pairs), which is proportional to the orientation factor $\kappa 2$, $J$ is the donor–acceptor spectral overlap in $M^{-1}\ cm^{-1}\ n10$, and the donor quantum yield $\Phi D$:

$$R06 = 8.79 \times 1023\kappa 2\Phi D Jn4 \ (in\ Å6)$$

Donor emission and acceptor absorption dipoles rotate fast in comparison to the donor fluorescence lifetime, and the $\kappa 2$ value can be replaced by its average value $\langle \kappa 2 \rangle$, which is equivalent to 2/3 isotropic rotation. The equation below provides a time dependent value of $\kappa 2$ where $\theta T$ represents the angle between the donor emission dipole and the acceptor absorption dipole and $\theta D$ (respectively, $\theta A$) exhibits the angle between the donor–acceptor connection line and the donor emission dipole (respectively, acceptor absorption dipole).

$$\kappa 2 = (\cos\theta T - 3\cos\theta D\cos\theta A)\ 2$$

The spectral overlap integral $J$ takes into consideration the normalized fluorescence emission spectrum of the donor $fD$ [$\int fD(\lambda)\ d\lambda = 1$] and the molar extinction coefficient of the acceptor $\varepsilon A$ (in $M^{-1}\ cm^{-1}$).

$$J = \int 0\infty fD\ (\lambda)\varepsilon A(\lambda)\lambda 4 d\lambda$$

The decrease in the lifetime of the donor is used to determine the FRET according to the following equation:

$$\tau = 1kr + kISC + kbl + k\ (FR)ET = \tau 0/(1 + k(FR)ET\tau 0)$$

Further, the FRET efficiency $E$ is calculated using:

$$E = kFRETkr + konr + kFRET = [1 + (RR0)6] - 1 = 1 - \tau\tau 0$$

or ratiometrically from the donor and acceptor fluorescence intensities FD and FA [12]:

$$E = FAFA + \gamma FD$$

## 3.7 Conclusion

Some of the applications of fluorescence spectroscopy are covered in this chapter with the help of salient examples that enable the user to derive knowledge for designing their own experiments. Also, the fundamental concepts and mathematical basics of fluorescence polarization and FRET are covered in this chapter. However, there are different dimensions to fluorescence spectroscopy that are beyond the scope of this chapter and therefore it is suggested to consult resources dedicated to individual domains such as time resolved fluorescence, fluorescence microscopy, etc.

## Acknowledgment

Author acknowledges the contribution of the students Anshika Gupta, Prachi Bhatia, and Priyansh Srivastava in literature collection and image preparation.

## References

[1] Albrecht C, Lakowicz JR. Principles of fluorescence spectroscopy. Anal Bioanal Chem 2008;390 (5):1223–4.

[2] Emptage NJ. Fluorescent imaging in living systems. Curr Opin Pharmacol 2001;1(5):521–5.

[3] Wilson K, Walker J, editors. Principles and techniques of practical biochemistry. Cambridge University Press; 2000.

[4] Albani JR. Structure and dynamics of macromolecules: absorption and fluorescence studies. Elsevier; 20040-444-51449-X. p. 58.

[5] Grum F. Instrumentation in fluorescence measurements. J Color Appearance 1972;18–27.

[6] Dong B, Almassalha LM, Stypula-Cyrus Y, Urban BE, Chandler JE, Sun C, et al. Superresolution intrinsic fluorescence imaging of chromatin utilizing native, unmodified nucleic acids for contrast. Proc Natl Acad Sci 2016;113 (35):9716–21.

[7] Misra G, Ramachandran R. Hsp70-1 from *Plasmodium falciparum*: protein stability, domain analysis and chaperone activity. Biophys Chem 2009;142(1-3):55–64.

[8] Misra G, Ramachandran R. Exploring the positional importance of aromatic residues and lysine in the interactions of peptides with the *Plasmodium falciparum* Hsp70-1. Biochim Biophys Acta (BBA)-Prot Proteom 2010; 1804(11):2146–52.

[9] Roy GB, et al. Chiral Salarg and its metal complex: unique extrinsic fluorophores. Spectrochim Acta Part A: Mol Biomol Spectrosc 2011;79(3):423–7.

[10] Gök E, Olgaz S. Binding of fluorescein isothiocyanate to insulin: a fluorimetric labeling study. J Fluorescence 2004;14(2):203–6.

[11] Zehentbauer FM, Moretto C, Stephen R, Thevar T, Gilchrist JR, Pokrajac D, et al. Fluorescence spectroscopy of Rhodamine 6G: concentration and solvent effects. Spectrochim Acta Part A: Mol Biomole Spectrosc 2014;121: 147–51.

[12] Michalet X, Weiss S, Jäger M. Single-molecule fluorescence studies of protein folding and conformational dynamics. Chem Rev 2006;106(5):1785–813.

# Chapter | 4 |

# High-throughput sequencing

Dibyabhaba Pradhan[1], Amit Kumar[1], Harpreet Singh[1] and Usha Agrawal[2]

[1]ICMR-AIIMS Computational Genomics Centre, Convergence Block, All India Institute of Medical Sciences, New Delhi, India, [2]CRIB Lab, ICMR-National Institute of Pathology, New Delhi, India

## 4.1 Introduction

High-throughput sequencing (HTS) describes modern sequencing technologies (Illumina, Ion Torrent, Pacific Biosciences, and Oxford Nanopore Technologies, etc.) developed in the new millennium. They are also referred as next generation, third generation, and fourth generation sequencing methods. HTS is fast, cheaper than traditional DNA sequencing methods, and generates massive amount of data. The present HTS technologies can sequence the human genome at a cost as low as $1121 in 2 days as opposed to $0.5−1 billion and over a decade's time. Third generation sequencing such as PacBio, produces longer raw reads than second generation sequencing methods [1]. Fourth generation next generation sequencing (NGS) technologies have potential to sequence the human genome for <$1000 [2,3]. Portability, affordability, and speed in data generation by MinION, released by Oxford Nanopore Technologies, has generated much interest in the genomics community [4]. Extraordinary advancements in HTS technologies in the last decade have shifted our path from the nascent era of the personal genome to practice genomic variation and expression in clinical diagnosis and precision medicine.

HTS, in general, involves genomics (DNA-seq), transcriptomics (RNA-seq), and epigenetic applications (ChIP-seq, Bisulfite-seq, MeDIP-seq). DNA-seq is achieved through whole-genome sequencing (WGS) and whole-exome sequencing (WES) to identify genomic variants (single nucleotide variants, insertions or deletions, copy number alternations, and structural variations) from biological samples [5]. RNA-seq is being used for quantification of gene expression profiles, detecting novel transcripts, alternative splicing, fusion transcripts, and RNA editing [6]. ChIP-seq involves chromatin immuno-precipitation (ChIP) and subsequent sequencing (Seq). It enables identification of transcription factors (TFs) as well as other protein binding sites in DNA [7−9]. Bisulfite-seq detects genome wide DNA methylation pattern [10], while MeDIP-seq is a cost-effective methylome alternative [11]. Systematic analysis of different kinds of HTS data and their integration is vital in achieving favorable clinical implications. Herein, an overview of HTS raw data preprocessing and data analysis protocols with examples have been discussed.

## 4.2 High-throughput sequencing raw data

NGS machines generates millions of short DNA fragments of 40−300 base length (typically <100 bases) from biological samples, called reads [12,13]. HTS reads may be single or pair end depending on the choice of library preparation during sequencing. Pair end reads (Fig. 4.1A) are more useful in achieving convincing results during downstream analysis, particularly in detection of gene fusions, genomic rearrangements, repetitive sequence elements, and novel transcripts. The raw reads are commonly stored in compressed FASTQ format [14]. Each read of FASTQ file comprises of four lines: (1) a sequence identifier starting with "@" followed by (2) read sequence, (3) a spacer denoted by " + " symbol, and (4) quality score. Each base call of a read is assigned with an ASCII value, referred to

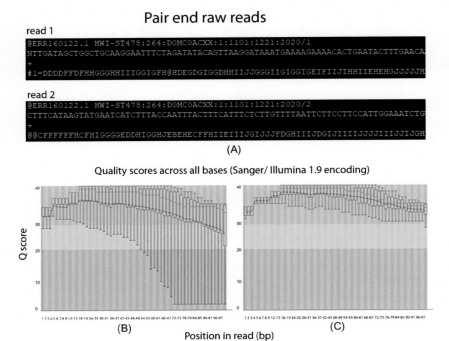

**Figure 4.1** Raw data and QC report (A) Raw read; Q score vs. read position (B) raw read 1 (C) QC filtered read 1 ($\geq 30$).

as Phred quality scores ($Q$) (Fig. 4.1A). The quality score encoding in NGS reads could be either Phred33 or Phred64 depending upon the version of sequencer used. The minimum quality score for a sample should be at least 20 whereas reads with $Q \geq 30$ are recommended for downstream analysis. $Q$ score of 30 represents $<0.001$ probability of erroneous base call.

## 4.3 Databases for storing next generation sequencing raw data

The European Nucleotide Archive (ENA) [15], NCBI Sequence Read Archive (SRA) [16], and DDBJ Sequence Read Archive (DRA) [17] are the three archive databases that store output data generated from HTS platforms through the International Nucleotide Sequence Database Collaboration (INSDC). Alongside, there are disease specific databases such as The Cancer Genome Atlas (TCGA), which catalog major cancer-causing genomic alterations [18,19]. The raw data along with sample, experimental setup, platform, and interpreted information available to the research community through these databases enhance

reproducibility and allow for new discoveries by comparing or merging data sets of interest [20].

## 4.4 Data processing

The HTS data processing involves quality control of raw reads followed by DNA-seq/RNA-seq/epigenetic analysis depending upon the experimental design. The final step involves filtering and annotation of the processed results and its clinical/biological interpretations.

### 4.4.1 Read quality control

Quality control is the first step of processing raw sequencing data that provides insight on quality of HTS reads. Base calling errors, poor quality reads/bases, and primer/adaptor contamination may lead to erroneous outcomes in downstream sequence analysis [21]. Quality control ensures filtering/trimming of raw reads to select high quality data with desired $Q$ score ($\geq 30$) and no contaminants or sequencing artifacts. QC and filtering of high-quality sequencing data at the end-user level is advisable prior to downstream analysis [22].

*Tools*

The widely used software for QC of raw reads includes FastQC, FastQ Screen, NGS QC Toolkit, FASTX-Toolkit, QC-Chain, RRINSEQ, and ClinQC [22,23]. FastQC is the most popular and can be used with data generated from all kinds of NGS platforms. It can be used with platform specific adaptor trimming software viz. Trimmomatic, Cutadapt, or fastx trimmer to retain high quality reads. ClinQC is a powerful and easy to handle pipeline that uses FASTQC, PRINSEQ, Alientrimme, and TraceTuner for quality control and trimming in clinical research [23].

---

### Example

*Sample*: The sample dataset with European Nucleotide Achieve project identifier PRJEB3132 is used for raw data processing exercises discussed in this study for quality checking, DNA-seq, and RNA-seq examples. The study encompasses both exome and transcriptome fastq files from lung adenocarcinomas and their adjacent normal lung tissues.

*Platform*: It is recommended to use Linux Operating System (18.04 LTS 64 bit Ubuntu).

*Task 1: Quality checking and Trimming*

*Prerequisites Tools:* FastQC [24] and Trimmomatic [25] need to be installed for the exercise.

*Steps*

- *Download raw reads for a single sample from project ID PRJEB3132*

    Open Linux terminal
    mkdir exome # create a directory name exome
    cd exome # Move into exome directory
    wget ftp://ftp.sra.ebi.ac.uk/vol1/fastq/ERR160/
    ERR160121/ERR160122_1.fastq.gz
    wget ftp://ftp.sra.ebi.ac.uk/vol1/fastq/ERR160/
    ERR160121/ERR160122_2.fastq.gz
    gunzip ERR160122_1.fastq.gz # output:
    ERR160122_1.fastq
    gunzip ERR160122_2.fastq.gz # output:
    ERR160122_2.fastq
    less ERR160122_1.fastq # Visualize contents of a fastq file of 1st read in terminal (Fig. 4.1A)

- *Check read quality*

    fastqc ERR160122_1.fastq # generates quality report for first raw reads (Fig. 4.1B)
    fastqc ERR160122_2.fastq # generates quality report for second raw reads

- *Trimming*

    java -jar trimmomatic-0.38.jar PE -phred33/
    ERR160122_1.fastq.gz ERR160122_2.fastq.gz output_
    ERR160122_1P.fastq.gz output_ ERR160122_1UP.fastq.gz

    *(Continued)*

---

### (cont'd)

    output_ERR160122_2P.fastq.gz output_ ERR160122_2UP.
    fastq.gz ILLUMINACLIP:adapters/TruSeq3-PE.fa:2:30:10
    LEADING:3 TRAILING:3 SLIDINGWINDOW:4:15 MINLEN:50
    # trims Illumina adapters, if any, and filters reads with Q score < 30 (Fig. 4.1C). In trimmomatic output P stands for paired, UP stands for unpaired reads.

*Note: "#" is used to mention comment lines.*

*QC report*

Quality check with FastQC on raw reads showed an average Q score ≥ 20. Reads with Q score ≥ 30 are filtered from raw reads using Trimmomatic. The filtered reads are suitable for downstream analysis.

## 4.4.2 DNA-seq: the next generation sequencing genomics application

DNA variation in a human genome comprises small base substitution in the form of single nucleotide variants (SNVs), insertions and deletions (indels), as well as structural variants due to large genomic amplifications/deletions (copy number alterations) and rearrangements such as inversions and translocations. DNA-seq could harvest full spectrum of these genomic variations in a single experiment as well as guide reference based or de novo assembly of the draft genome. Moreover, DNA-seq analysis is applicable to whole genome, whole exome, and clinical exome data in both case—controls as well as familial studies (trios analysis) at a population scale.

### 4.4.2.1 Short variant discovery

Short variant discovery (≤10 bp) from quality filtered NGS reads typically involves mapping and alignment of reads with reference and variant calling [26]. Mapping and alignment starts with indexing of reads and reference sequences, which saves both time and memory by aiding the aligner to zero in potential origin of a read within the genome. Reads mapped to the reference are documented in a sequence alignment map (SAM) file, which contains quality score, mapped genomic location, mapping confidence, mate read information, and a CIGAR string outlining SNPs contained in the read. SAM files are stored in compressed binary versions as BAM (binary sequence alignment map) files. The alignment files are evaluated for several quality metrics viz. average or median depth, coverage, mapping quality, insert size, and the number of discordantly mapped pairs. During this step, PCR deduplication and base quality recalibration of the deduplicated BAM file using known variants are also

**41**

recommended. Subsequently, short raw variants are called from the BAM file using variant caller algorithms [27].

The short variants are universally stored in variant calling format (VCF), which contains information about genomic location of variant with nucleotide base in reference allele and alternate allele, quality score, and genotype information. Probabilistic modeling along with a set of quality filters viz. quality score, mapping quality, quality by depth, etc. is applied to distinguish real variants from sequencing artifacts [22]. Overall quality of variant files are evaluated based on transition/transversion (Ti/Tv) ratio, number of novel nonsynonymous SNPs, genotype consistency between SNP chip data and exome sequencing data, the heterozygosity to nonreference homozygosity ratio, etc. [22]. Furthermore, variants are annotated with known short variant databases such as dbSNP [28], ClinVar [29,30], 1000 Genome [31], ESP6500, ExAC, UK10K [32], and HapMap [33]; and annotation tools like snpEff [34], ANNOVAR [35], etc. to filter variants with known clinical significance as well as novel ones.

## 4.4.2.2 Structural variant discovery

Genomic structural variations play an important role in phenotypic diversity and disease susceptibility. DNA-seq analysis involving read-pair, split-read, read-depth, and assembly algorithms can pick up structural variants from reads aligned to reference genome. Read-pair method uses insert size data from sequencing library and evaluates every abnormal region in the aligned reads to detect structural variants. Algorithm such as split-read breaks short reads into $\geq 2$ fragments before mapping them individually to the reference genome. The mapping location and orientation of the split-mapped reads aid detection of small and medium-sized structural variants. Relative copy numbers (gains/loss) can be estimated from significantly higher or lower short read mapping density (read depth) in a genomic region by assuming a random distribution of mapping depth. Larger ($>1$ kb) copy number variants (CNVs) can be effectively identified by these methods. Structural variants are also detected through reference based local assembly of genomic locations. Several tools implement all four algorithms to identify structural variants. Moreover, advent of single molecules sequencing technologies would improve SV detection accuracy.

*Software and pipelines*

Some of the commonly used software and pipelines used for short and structural variant discovery have been listed in Table 4.1. BWA [36] and Bowtie2 [37] are the primary choice for indexing and alignment of reads to human genome, while, most of the pipelines for variant discovery implements Genome Analysis Tool Kit (GATK) [38] for variant calling. Similarly, filtering and annotations of variant files are achieved through snpEff and ANNOVAR. These two softwares implement databases (dbSNP, ClinVar, 1000 Genome, ESP6500, and HapMap) and tools (SIFT [39], PolyPhen-2 [40]) for predicting functional consequences of mutations. In addition, human genome mutation database (HGMD) [41] is also used for effective annotation of variants. Among pipelines, SeqMule, Impact, DNAP are the well-known ones for variant discovery [42]. These pipelines implement variant callers and annotation tools as well as databases systematically for automated variant discovery. Moreover, exomesuite, mirTrios, etc. have been developed for familial/trios analysis.

GATK best practice workflow was implemented for variant calling in a metabolomics genome wide association study (mGWAS) to uncover 21 common variant and 12 functional rare variant mQTLs (methylation quantitative trait loci), of which 45% are novel. These variants would be having high implication for precision medicine in the Middle East [43]. Similarly, GATK, BWA, and MuTect used together to elucidate the genomic architecture of Asian EGFR-mutant lung adenocarcinoma [44].

Clinical actionable variants are also identified by Dewey et al. implementing GATK best practice pipeline from combined exome sequencing results for over 50,000 people with their electronic health records [59]. Jin et al. deciphered a de novo mutation in BSG gene among early-onset high myopia patients by analyzing whole exome reads of 18 trios [60]. Guo et al. implemented STRiP, Delly, Manta, BreakDancer, and Pindel tools to identify structural variants in TP53, PALB2, PTEN, and RAD51C genes that would be of interest for early clinical diagnosis of breast cancer [61]. These studies advocate promises of exome sequencing and their integration with electronic health records in precision medicine. A standard variant calling workflow implementing tools/pipelines from whole genome/exome/clinical exome reads of human sample is given as Fig. 4.2.

## 4.4.2.3 Genome assembly

The primary application of DNA sequencing experiments is constructing a draft genome of an organism. The large number of DNA-seq reads generated from NGS experiments can be used in assembling a population based or strain specific draft genome. Genome assembly can be achieved through reference based or in a de novo manner [62]. The reference based approach follows alignment of raw reads to existing reference and constructing population specific genome based on observed genetic variations and by incorporating high quality variants detected through the variant callers. Such approaches aid detecting minor allele frequencies of a gene across population as well as enhance our understanding of genetic diversity in

**Table 4.1 Tools and pipelines for whole genome/exome data analysis.**

| Program | Task | Ref. |
| --- | --- | --- |
| Burrows—Wheeler Transform (BWA): BWA-backtrack, BWA-SW and BWA-MEM | Indexing and alignment—Mapping low-divergent sequences against a large reference genome. BWA-MEM is latest and more accurate; support longer sequences (70 bp to 1 Mbp) gapped and split alignment. | [36,45] |
| Bowtie2 | Indexing and alignment—mapping reads (50—1000 bp) to large reference genome with small memory footprint. Supports gapped alignment. | [37] |
| Samtools | SNP calling—Utilities for indexing genome and reads, sorting, merging, mutation calling. | [46] |
| Genome Analysis Tool Kit (GATK) | SNV calling—Preprocessing of alignment files (SAM/BAM), variant discovery and genotyping—supports both germline (HaplotypeCaller) and somatic variant calling (MuTect2). | [38,47,48] |
| SNVSniffer | SNV calling—Supports germline and somatic variant calling. | [49] |
| SVDetect | SV detection—Combines RC and RP algorithms. | [50] |
| Genome STRiP | SV detection—Combines RP, RC, SR, and population-scale patterns. | [51] |
| PRISM | SV detection—Implements SR and RC/RP for SV detection. | [52] |
| DELLY | SV detection—Suitable for CNV detection, inversions or reciprocal translocations. | [53] |
| LUMPY | SV detection—Suitable with low coverage data or low variant allele frequency. | [54] |
| GATK Best Practice pipeline | FastQ data to high confidence variant calls. | [48] |
| SeqMule | Analysis pipeline for exome/genome HTS data. | [55] |
| IMPACT | Pipeline to detect clinically actionable variants from whole exome samples. | [56] |
| DNAp | A pipeline for DNA-seq data analysis. | [57] |
| COBASI | De novo SNV calling pipeline—Precise detection of de novo single nucleotide variants in human genomes. | [58] |

world populations across ethnicities that in turn is used to filter rare genetic variants of clinical interest. Reference based assemblies are also used for constructing pan genomes for multiple strains of infectious pathogens. The conserved genomic locations in the pan genome with immunogenic properties are used as potent multisubunit vaccine candidates against the infectious pathogen group. The software listed in whole exome data analysis can be used effectively for reference based genome assembly.

De novo assembly technique is being used by researchers to assemble the genome of an organism for which a draft genome is not available. Raw reads are mapped against each other to identify overlap between them to build larger contiguous sequences called contigs. Ordered contigs are joined together by padding character "N" to an even larger genomic fragment known as a scaffold. Pair end reads and mate pair reads are useful to join reads/contigs at a distance of 5—10 kb apart. Accordingly, scaffolds are joined into linkage groups to construct a draft genome and annotated through sequence alignment with the genome of a homologous organism. Expected genome size and N50 statistics, which specifies contigs representing 50% of the assembled nucleotide, act as indicators of genome assembly quality [63]. However, it is recommended to use multiple genome assembly software iteratively to construct draft genome of an organism.

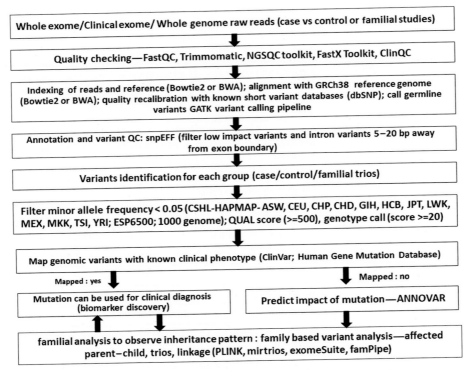

**Figure 4.2** Standard variant calling workflow implementing tools/pipelines from whole genome/exome/clinical exome reads of human sample.

Popular genome assembly software includes Velvet [64], SPAdes [65], SOAPdenovo [66], and ABySS [67]. Most of them are developed based on De Bruijn graph algorithms. Velvet and SPAdes are more effective in assembly of draft genome while ABySS is better at assembling larger genomes. The genome assemblies can be validated through QUAST-LG tool [68]. Sellera et al. reported presence of genes conferring drug resistance to aminoglycosides, β-lactams, phenicols, sulfonamides, tetracyclines, quinolones, and fosfomycin through de novo assembly of drug-resistant *Pseudomonas aeruginosa* genome using Velvet [69]. Similarly, de novo assembly and annotation of *Vibrio campbellii* RT-1 strain is reported based on analysis through Velvet and SPAdes [70].

---

**Examples**

*Sample*: The sample dataset for consists of whole exome raw reads from lung cancer tissue (*n* = 2) available at European nucleotide Achieve project identifier PRJEB3132.

*Platform*: It is recommended to use Linux Operating System (18.04 LTS 64 bit Ubuntu).

*(Continued)*

---

(cont'd)

*Task 2: Short Variant Discovery*

*Prerequisites Tools:* Install Genome Analysis Tool Kit [71], Burrows—Wheeler Aligner [72], and their dependencies.

```
#Download two lung adenocarcinoma samples
mkdir exome
cd exome
Download pair end reads for sample ERR166338 and ERR166339.
gunzip *.gz # extract fastq files
fastqc *.fastq # Generate Q report and filter reads based on Qscore and trimming, if required.
# Index reference genome
wget ftp://ftp.ensembl.org/pub/release-93/fasta/homo_sapiens/dna/Homo_sapiens.GRCh38.dna.primary_assembly.fa.gz # Download GRCh38 version of human genome
gunzip Homo_sapiens.GRCh38.dna.primary_assembly.fa.gz
mv Homo_sapiens.GRCh38.dna.primary_assembly.fa genome.fa
bwa index genome.fa
```

*(Continued)*

(cont'd)

```
samtools faidx genome.fa
    gatk CreateSequenceDictionary --REFERENCE = genome.
fa --OUTPUT = genome.dict
    # Alignment of reads with reference
    bwa mem genome.fa ERR166338_1.fastq
ERR166338_2.fastq > ERR166338.sam
    gatk SortSam -I = ERR166338.sam -O = ERR166338.bam
--SORT_ORDER = coordinate
    # Alignment QC
    gatk CollectAlignmentSummaryMetrics —R = genome.fa
-I = ERR166338.bam -O = alignment_metrics.txt
--VALIDATION_STRINGENCY = LENIENT
    gatk CollectInsertSizeMetrics -I = ERR166338.bam
-O = insert_metrics.txt
--Histogram_FILE = insert_size_histogram.pdf
    gatk AddOrReplaceReadGroups -I = ERR166338.bam
-O = ERR166338_readgroup.bam --RGLB = lib1
--RGPL = illumina --RGPU = NONE --RGSM = BG
    # Variants calling
    gatk MarkDuplicates -I = ERR166338_readgroup.bam
-O = ERR166338_dedup_reads.bam
--METRICS_FILE = metrics.txt -AS = true
--VALIDATION_STRINGENCY = LENIENT
    gatk BuildBamIndex -I = ERR166338_dedup_reads.bam
    gatk BaseRecalibrator -R genome.fa -I
ERR166338_dedup_reads.bam --known-sites
All_20180418.vcf -O recal_data.table # Note contigs
should be converted to chromosome in genome.fa to
match dbSNP, if applicable using sed command.
    gatk ApplyBQSR -R genome.fa -I
ERR166338_dedup_reads.bam -bqsr recal_data.table -O
ERR166338_recal_reads.bam
    gatk HaplotypeCaller --dbsnp All_20180418.vcf -R
genome.fa -I ERR166338_recal_reads.bam -O
raw_variants_ERR166338.g.vcf -ERC GVCF
    # similarly, generate "raw_variants_ERR166339.g.vcf" for
sample ERR166339 from its raw reads.
    # Joint variant call from both samples
    gatk CombineGVCFs --dbsnp All_20180418.vcf -R
genome.fa --variant raw_variants_ERR166339.g.vcf
--variant raw_variants_ERR166338.g.vcf -O lcancer.vcf
    # Genotype call
    gatk GenotypeGVCFs --dbsnp All_20180418.vcf -R
genome.fna -V lcancer.vcf -O lcancer_genotype.vcf
    #Variant Annotation and filtering—Annovar
    ./table_annovar.pl lcancer_genotype.vcf humandb/
-buildver hg38 -out vcf_myanno -remove -protocol
refGene,cytoBand,exac03,avsnp150,dbnsfp30a -operation
g,r,f,f,f -nastring. —vcfinput
    Results
    There are 614,584 short variants in vcf file generated
after joint variant call and genotype call (Suppl. data
```

(Continued)

(cont'd)

1_DNAseq: lcancer_genotype.vcf). The vcf file further annotated and filtered for synonymous variants. The resulting filtered variant file showed 778 deleterious mutations (Suppl. data 1_DNAseq: polyphen2-bothDB.txt) and 634 possible/probable damaging mutations (Suppl. data 1_DNAseq: SIFT-pred.txt). Moreover, the joint variant call file after genotyping is provided as supplementary material (Suppl. data 1_DNAseq: lcancer_genotype.vcf) for users to apply filter for QUAL score, minor allele frequency, genotype quality as well as mapping with variant databases such as ClinVar to annotate known variants linked to a clinical phenotype.

## 4.4.3 RNA-seq: the next generation sequencing transcriptomics application

RNA-seq quantifies expression of entire transcriptome, identifies novel transcript, gene fusions, and RNA editing in a given experimental condition. Specialized types of RNA-seq experiments are also being reported to analyze expression of noncoding RNAs in disease phenotypes. In RNA-seq, reads need to map with exons coding for transcripts, therefore, the mapping programs are designed to recognize exon—intron boundaries and align with exons only. Differential gene expression analysis implementing RNA-seq protocols has been immensely successful in identifying potential drug targets and biomarkers for several diseases of interest.

### 4.4.3.1 Differential gene expression analysis

The RNA-seq reads need to be quality filtered, indexed, and aligned using RNA-seq specific aligners. Abundance of each transcript is computed and normalized across samples to represent FPKM (fragments per kilobase of transcript per million mapped reads) or RPKM (reads per kilobase per million mapped reads). Differential expression among two or more than two experimental conditions are calculated by applying statistical test of significance—moderated T-test or one way ANOVA—attached to specialized tools for gene expression analysis [73]. The differentially expressed genes are further annotated with gene ontology, pathway enrichment, and gene network analysis to identifying differentially expressed genes with clinical significance (Fig. 4.3).

Beyond gene expression analysis, RNA-Seq can also be applied to discover novel gene structures, alternatively spliced isoforms, allele-specific expression, and even for variant discovery. However, utility of RNA-seq variants

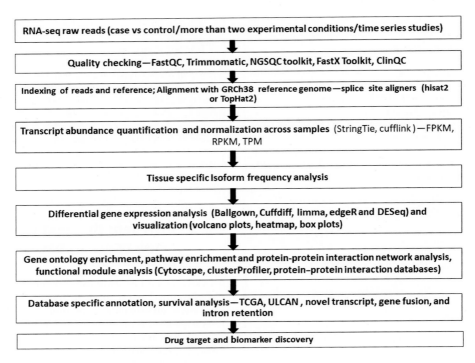

**Figure 4.3** Standard protocol for analysis of RNA-seq reads.

becomes challenging with downregulated genes, reverse transcription error, RNA editing site, and splice-site alignment error rate. In cases where both DNA-seq and RNA-seq reads are available, variants called from RNA-seq reads could be used to evaluate variants detected from DNA-seq experiments.

### 4.4.3.2 Transcriptome reconstruction

As that of genome assembly, RNA-seq reads can also be assembled to reconstruct whole transcriptome of an organism based on a reference guided approach or de novo manner. Reference guided approach follows standard indexing, alignment, and transcript abundance quantification methods as explained in the differential gene expression section. The de novo assembly approach builds consensus transcripts from short RNA-seq reads without a reference. De novo transcriptome construction is useful where reference genome/transcriptome of an organism is yet unknown.

*Software and pipelines*

TopHat2 [74] or hisat2 [75] are the software programs commonly used for indexing and alignment of reference

genome with RNA-seq reads. Quantification of genes and normalization across samples are carried out using Cufflink [76] or Stringtie [75]. The differential gene expression analyses of quantified genes are carried out using CuffDiff [76,77] or Ballgown [75]. Other tools used for RNA-seq analysis include DEseq2 [78], edgeR [79], and limma [80]. SARTools is a DEseq2 and edgeR based pipeline for RNA-seq data analysis [81]. Pathway analysis and gene ontology enrichment can be performed using clusterProfiler [82] or DAVID webserver [83]. Network analysis could be performed using protein–protein interaction databases such as STRING or Biogrid through Cytoscape interface, a tool primarily used for analysis and visualization of the protein–protein interaction networks [84,85]. Trinity [86] is the commonly used for de novo transcriptome reconstruction. Kunz et al. implemented RNA-seq based transcriptome data analytic techniques on melanocytic nevi and primary melanoma samples to identify gene signatures underlying in early and late stage melanoma with relevance for diagnosis and therapy [87]. Similarly, Arun et al. reported integrin drug targets for 17 solid tumor types by analyzing gene expression data from TCGA [88].

## Examples

*Sample*: The sample dataset consists of whole transcriptome raw reads from lung cancer tissue ($n = 2$) and adjacent normal lung tissues ($n = 4$) available at European nucleotide Achieve project identifier PRJEB3132.

*Platform*: It is recommended to use Linux Operating System (18.04 LTS 64 bit Ubuntu).

*Task 3: Differential Gene Expression Analysis*

*Prerequisites Tools*: The following software along with their dependency packages need to installed for executing the given exercise. Detailed descriptions for installation are explained by [75].

HISAT2 [89], StringTie [90], SAMtools [91], R [92], Bioconductor, ballgown with dependencies [93], and Cytoscape [94]

*Steps*

- *Download raw reads of six samples*
  mkdir transcriptome # create a directory name transcriptome
  cd transcriptome # Move into transcriptome directory
  mkdir data # create directories to place samples, genome, gene annotation files and indexes
  cd data
  mkdir samples
  mkdir indexes
  mkdir genes
  mkdir genome
  cd samples
  ./ transcriptome_download.sh # download six samples mentioned in the script (*Suppl. data 2_RNAseq: transcriptome_download.sh*)
- *Check read quality*
  fastqc *.gz
  Use trimmomatic as described in QC section, if trimming is required.
- *Indexing human genome*
  cd./genome
  cp././exome/genome.fa./ # copy *genome.fa*—the hg38 version of human genome downloaded for Task 2 —into the genome directory.
  cd./genes
  wget ftp://ftp.ensembl.org/pub/release-93/gtf/homo_sapiens/Homo_sapiens.GRCh38.93.gtf.gz #Download corresponding GTF file
  gunzip Homo_sapiens.GRCh38.93.gtf.gz
  mv Homo_sapiens.GRCh38.93.gtf gene.gtf # Rename GTF as gene.gtf
  cd./indexes
  extract_splice_sites.py./genes/gene.gtf >gene.ss
  extract_exons.py./genes/ gene.gtf >gene.exon
  *(Continued)*

(cont'd)

  hisat2-build --ss gene.ss --exon gene.exon./genome/genome.fa genome_tran
- *Identifying differentially expressed genes*
  cd././# change to transcriptome directory. The files such as conditions.csv (*Suppl. data 2_RNAseq*), rnaseq_pipeline.config.sh (*Suppl. data 2_RNAseq*), rnaseq_pipeline.sh (*Suppl. data 2_RNAseq*) and rnaseq_ballgown.R (*Suppl. data 2_RNAseq*) should be placed in the present directory—these files generated by modifying rna_seq pipeline provided by [75] to execute the present examples.
  ./rnaseq_pipeline_config.sh # Check if rnaseq pipeline pre-requisites have been fulfilled.
  ./rnaseq_pipeline.sh
  Two files with differentially expressed transcripts and genes would be generated (*suppl. data 6, suppl. data 7*).
- *Downstream analysis*: Generate network in Cytoscape using String App (Fig. 4.4) [84].

*Results*

Seventy-five transcripts of known genes with q value $< 0.05$ and fold change $\pm 2$ are deemed as differentially expressed transcripts (Fig. 4.4, *Suppl. data 2_RNAseq: transcript.csv*). IGHJ3, IFITM1, SET, SEL1L3, and IGKV2D-40 were observed to be top five upregulated genes while SMARCA2, SMARCD2, WDR6, ZNF146, and TK2 were identified to be the top five downregulated genes. Such dysregulated expression of IFITM1, SET, SMARCA2, SMARCD2 are already known to be associated with cancer [95–97]. In case of gene SNX6, one transcript is highly upregulated while another is highly downregulated, such contentious genes are excluded from further downstream analysis in this example, however, in such scenario, the researcher may either look at biological context, tissue type to make out which transcript is most likely to express or compute mean/median/mode from the FPKM values to conclude if the gene is upregulated or downregulated. Network and module analysis have revealed that PXN, SMARCA2, G3BP1, and MYO9B are the genes of clinical interest as potential drug target or biomarkers. The study on the example dataset has reciprocated previous reports showing a critical role of these genes in tumor progression and metastasis [76,78–80].

## 4.4.4 Next generation sequencing epigenetic application

Epigenetic modifications play an important role in contributing to disease severity. NGS based epigenetic application strives to distinguish genome wide epigenetic marks through Bisulfite-seq and ChIP-seq applications.

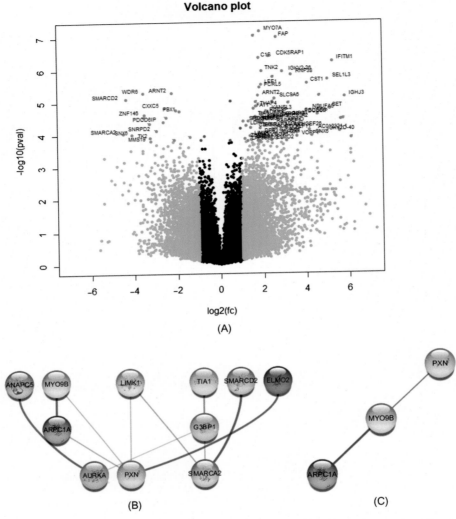

**Figure 4.4** (A) Differentially expressed transcripts between lung cancer and its adjacent normal tissues: Significantly up and downregulated transcripts are showed in green dots. Gene symbols are labelled for transcripts wherever annotation available. (B) PXN, SMARCA2 and G3BP1 are identified as hub genes. (C) MYO9B is identified as seed gene in module analysis.

Identification of genome wide epigenetic markers viz. DNA methylation and protein/DNA binding landscapes among individuals could be indicators of their life style.

*Tools*

Bicycle pipeline is an effective method to analyze Bisulfite-seq data [98] while ChIP-Seq data could be analyzed using a customized pipeline of Bioconductor packages dada2, QuasR, mosaics, and ChIPseeker [99]. Both pipelines accept fastq files as input and report genome wide DNA methylation pattern and DNA/protein interactions, respectively.

## 4.5 High-throughput sequencing data analysis in an open source platformgalaxy

Galaxy is an open source, user friendly, web-based platform for analyzing HTS DNA-seq, RNA-seq, and Chip-seq data. Users need to register at https://galaxyproject.org, then upload raw data and start analysis from quality control to the desired HTS application. Galaxy platform can also be set up in offline mode in the laboratory to facilitate data analysis. In offline mode, the user needs to install the required applications and workflows from Tool Shed to perform data analysis [100].

## 4.6 Conclusion

High-throughput genomic technologies have been moving forward in the way of implementing clinical genomics in patient care. Whole exome sequencing has been explored to generate diagnosis reports with clinically actionable variants. At this juncture, systematic analysis of HTS data is critical to realize the goal of precision medicine. The comprehensive list of tools and pipelines along with data analysis protocols discussed in this chapter would provide brief insight to researchers on HTS data analysis. Moreover, the Galaxy platform could be an ideal step to start with for a biologist naïve to the Linux operating

system. It must be noted that commercial software such as DNASTAR, CLC Genomics provides other alternatives for HTS data analysis. These software packages are user friendly, however, they require a commercial license. Nevertheless, commercial software also uses open source NGS tools for HTS data analysis with additional customization and graphics. To summarize, the increasing knowledgebase of genomic variants/biomarkers/drug targets due to HTS would definitely guide precision medicine through specific genetic diagnosis.

## Acknowledgment

Authors are grateful to Indian Council of Medical research for financial support through Computational Genomics Centre.

## Reference

[1] Rhoads A, Au KF. PacBio sequencing and its applications. Genomics Proteomics Bioinformatics 2015; 13(5):278–89.

[2] Deamer D, Akeson M, Branton D. Three decades of nanopore sequencing. Nat Biotechnol 2016; 34(5):518–24.

[3] Feng Y, Zhang Y, Ying C, Wang D, Du C. Nanopore-based fourth-generation DNA sequencing technology. Genomics Proteomics Bioinformatics 2015;13(1):4–16.

[4] Lu H, Giordano F, Ning Z. Oxford nanopore MinION sequencing and genome assembly. Genomics Proteomics Bioinformatics 2016; 14(5):265–79.

[5] Shyr D, Liu Q. Next generation sequencing in cancer research and clinical application. Biol Proced Online 2013;15(1):4.

[6] Lappalainen T, Sammeth M, Friedländer MR, 't Hoen PAC, Monlong J, Rivas MA, et al. Transcriptome and genome sequencing uncovers functional variation in humans. Nature 2013;501(7468):506–11.

[7] Raha D, Hong M, Snyder M. ChIP-Seq: a method for global

identification of regulatory elements in the genome. Curr Protoc Mol Biol 2010;1–14 Chapter 21: Unit 21.19.

[8] Park PJ. ChIP-seq: advantages and challenges of a maturing technology. Nat Rev Genet 2009;10(10): 669–80.

[9] Kidder BL, Hu G, Zhao K. ChIP-Seq: technical considerations for obtaining high-quality data. Nat Immunol 2011;12(10):918–22.

[10] Li Y, Tollefsbol TO. DNA methylation detection: bisulfite genomic sequencing analysis. Methods Mol Biol 2011;791:11–21.

[11] Staunstrup NH, Starnawska A, Nyegaard M, Christiansen L, Nielsen AL, Børglum A, et al. Genome-wide DNA methylation profiling with MeDIP-seq using archived dried blood spots Clin Epigenetics 2016;8[cited 2018 Jun 27]. Available from: . Available from: https://www.ncbi.nlm.nih. gov/pmc/articles/PMC4960904/.

[12] Weitschek E, Santoni D, Fiscon G, De Cola MC, Bertolazzi P, Felici G. Next generation sequencing reads comparison with an alignment-free distance BMC Res

Notes 2014;7[cited 2018 Jul 11]. Available from: . Available from: https://www.ncbi.nlm.nih.gov/ pmc/articles/PMC4265526/.

[13] Muzzey D, Evans EA, Lieber C. Understanding the basics of NGS: from mechanism to variant calling. Curr Genet Med Rep 2015;3 (4):158–65.

[14] Numanagić I, Bonfield JK, Hach F, Voges J, Ostermann J, Alberti C, et al. Comparison of high-throughput sequencing data compression tools. Nat Methods 2016;13(12):1005–8.

[15] European Nucleotide Archive < EMBL-EBI. [cited 2018 Aug 26]. Available from: https://www.ebi. ac.uk/ena.

[16] Home - SRA - NCBI. [cited 2018 Aug 26]. Available from: https:// www.ncbi.nlm.nih.gov/sra.

[17] DDBJ Sequence Read Archive - Home. [cited 2018 Aug 26]. Available from: https://www. ddbj.nig.ac.jp/dra/index-e.html.

[18] Home - The Cancer Genome Atlas - Cancer Genome - TCGA. [cited 2018 Aug 26]. Available from: https://cancergenome.nih. gov/.

[19] Cancer Genome Atlas Research Network, Weinstein JN, Collisson EA, Mills GB, Shaw KRM, Ozenberger BA, et al. The cancer genome atlas pan-cancer analysis project. Nat Genet 2013;45(10):1113–20.

[20] Zhang Z, Li H, Jiang S, Li R, Li W, Chen H, et al. A survey and evaluation of Web-based tools/databases for variant analysis of TCGA data. Brief Bioinform 2018. Available from: https://doi.org/10.1093/bib/bby023.

[21] Khan NS, Saxena P, Pradhan D. NGS file formats and QC. Next Generation Sequencing Data Analysis: integrating genomics, transcriptomics and proteomics data for potential therapeutic target discovery. New Delhi: Biomedical Informatics Centre, ICMR-National Institute of Pathology; 2018. p. 21–6.

[22] Guo Y, Ye F, Sheng Q, Clark T, Samuels DC. Three-stage quality control strategies for DNA resequencing data. Brief Bioinform 2014;15(6):879–89.

[23] Pandey RV, Pabinger S, Kriegner A, Weinhäusel A. ClinQC: a tool for quality control and cleaning of Sanger and NGS data in clinical research. BMC Bioinformatics 2016;17:56.

[24] Babraham Bioinformatics - FastQC: a quality control tool for High Throughput Sequence Data. [cited 2018 Aug 26]. Available from: https://www.bioinformatics.babraham.ac.uk/projects/fastqc/.

[25] USADELLAB.org - Trimmomatic: a flexible read trimming tool for Illumina NGS data. [cited 2018 Aug 26]. Available from: http://www.usadellab.org/cms/?page = trimmomatic.

[26] Xu C. A review of somatic single nucleotide variant calling algorithms for next-generation sequencing data. Comput Struct Biotechnol J 2018;16:15–24.

[27] Verma R, Chaudhary S, Pradhan D. Alignment and variant calling. Next Generation Sequencing Data Analysis: integrating genomics, transcriptomics and proteomics data for potential therapeutic target discovery. New Delhi:

Biomedical Informatics Centre, ICMR-National Institute of Pathology 2018. p. 29–35.

[28] Kitts A, Phan L, Ward M, Holmes JB. The database of short genetic variation (dbSNP). National Center for Biotechnology Information (US) 2014[cited 2018 Jul 16]. Available from: https://www.ncbi.nlm.nih.gov/books/NBK174586/.

[29] Henrie A, Hemphill SE, Ruiz-Schultz N, Cushman B, DiStefano MT, Azzariti D, et al. ClinVar Miner: demonstrating utility of a Web-based tool for viewing and filtering ClinVar data. Hum Mutat 2018;39(8):1051–60.

[30] Toledo RA, NGS in PPGL (NGSnPPGL) Study Group. Inflated pathogenic variant profiles in the ClinVar database. Nat Rev Endocrinol 2018;14(7):387–9.

[31] 1000 Genomes Project Consortium, Auton A, Brooks LD, Durbin RM, Garrison EP, Kang HM, et al. A global reference for human genetic variation. Nature 2015;526(7571):68–74.

[32] Jiang Y, Li Z, Liu Z, Chen D, Wu W, Du Y, et al. mirDNMR: a gene-centered database of background de novo mutation rates in human. Nucl Acids Res 2017;45(D1):D796–803.

[33] Huang E, Liu C, Zheng J, Han X, Du W, Huang Y, et al. Genome-wide screen for universal individual identification SNPs based on the HapMap and 1000 Genomes databases. Sci Rep 2018;8(1):5553.

[34] Cingolani P, Platts A, Wang LL, Coon M, Nguyen T, Wang L, et al. A program for annotating and predicting the effects of single nucleotide polymorphisms, SnpEff: SNPs in the genome of Drosophila melanogaster strainw1118; iso-2; iso-3. Fly (Austin) 2012;6(2):80–92.

[35] Wang K, Li M, Hakonarson H. ANNOVAR: functional annotation of genetic variants from high-throughput sequencing data. Nucl Acids Res 2010;38(16):e164.

[36] Li H, Durbin R. Fast and accurate long-read alignment with Burrows-Wheeler transform. Bioinformatics 2010;26(5):589–95.

[37] Langmead B, Salzberg SL. Fast gapped-read alignment with Bowtie 2. Nat Methods 2012;9(4):357–9.

[38] McKenna A, Hanna M, Banks E, Sivachenko A, Cibulskis K, Kernytsky A, et al. The Genome Analysis Toolkit: a MapReduce framework for analyzing next-generation DNA sequencing data. Genome Res 2010;20(9):1297–303.

[39] Ng PC, Henikoff S. SIFT: predicting amino acid changes that affect protein function. Nucl Acids Res 2003;31(13):3812–14.

[40] Adzhubei IA, Schmidt S, Peshkin L, Ramensky VE, Gerasimova A, Bork P, et al. A method and server for predicting damaging missense mutations. Nat Methods 2010;7(4):248–9.

[41] Stenson PD, Mort M, Ball EV, Shaw K, Phillips A, Cooper DN. The Human Gene Mutation Database: building a comprehensive mutation repository for clinical and molecular genetics, diagnostic testing and personalized genomic medicine. Hum Genet 2014;133(1):1–9.

[42] Hwang S, Kim E, Lee I, Marcotte EM. Systematic comparison of variant calling pipelines using gold standard personal exome variants. Sci Rep 2015;5:17875.

[43] Yousri NA, Fakhro KA, Robay A, Rodriguez-Flores JL, Mohney RP, Zeriri H, et al. Whole-exome sequencing identifies common and rare variant metabolic QTLs in a Middle Eastern population. Nat Commun 2018;9(1):333.

[44] Nahar R, Zhai W, Zhang T, Takano A, Khng AJ, Lee YY, et al. Elucidating the genomic architecture of Asian EGFR-mutant lung adenocarcinoma through multiregion exome sequencing. Nat Commun 2018;9(1):216.

[45] Li H, Durbin R. Fast and accurate short read alignment with Burrows-Wheeler transform. Bioinformatics 2009;25(14):1754–60.

[46] Li H. A statistical framework for SNP calling, mutation discovery, association mapping and population genetical parameter

estimation from sequencing data. Bioinformatics 2011;27(21): 2987−93.

[47] DePristo MA, Banks E, Poplin R, Garimella KV, Maguire JR, Hartl C, et al. A framework for variation discovery and genotyping using next-generation DNA sequencing data. Nat Genet 2011;43(5): 491−8.

[48] Van der Auwera GA, Carneiro MO, Hartl C, Poplin R, Del Angel G, Levy-Moonshine A, et al. From FastQ data to high confidence variant calls: the Genome Analysis Toolkit best practices pipeline. Curr Protoc Bioinformatics 2013; 43(11.10):1−33.

[49] Liu Y, Loewer M, Aluru S, Schmidt B. SNVSniffer: an integrated caller for germline and somatic single-nucleotide and indel mutations. BMC Syst Biol 2016;10(Suppl. 2):47.

[50] Zeitouni B, Boeva V, Janoueix-Lerosey I, Loeillet S, Legoix-né P, Nicolas A, et al. SVDetect: a tool to identify genomic structural variations from paired-end and mate-pair sequencing data. Bioinformatics 2010;26(15): 1895−6.

[51] Handsaker RE, Korn JM, Nemesh J, McCarroll SA. Discovery and genotyping of genome structural polymorphism by sequencing on a population scale. Nat Genet 2011;43(3):269−76.

[52] Jiang Y, Wang Y, Brudno M. PRISM: pair-read informed split-read mapping for base-pair level detection of insertion, deletion and structural variants. Bioinformatics 2012;28(20): 2576−83.

[53] Rausch T, Zichner T, Schlattl A, Stütz AM, Benes V, Korbel JO. DELLY: structural variant discovery by integrated paired-end and split-read analysis. Bioinformatics 2012;28(18):i333−9.

[54] Layer RM, Chiang C, Quinlan AR, Hall IM. LUMPY: a probabilistic framework for structural variant discovery. Genome Biol 2014;15 (6):R84.

[55] Guo Y, Ding X, Shen Y, Lyon GJ, Wang K. SeqMule: automated

pipeline for analysis of human exome/genome sequencing data. Sci Rep 2015;5:14283.

[56] Hintzsche J, Kim J, Yadav V, Amato C, Robinson SE, Seelenfreund E, et al. IMPACT: a whole-exome sequencing analysis pipeline for integrating molecular profiles with actionable therapeutics in clinical samples. J Am Med Inform Assoc 2016;23(4):721−30.

[57] Causey JL, Ashby C, Walker K, Wang ZP, Yang M, Guan Y, et al. DNAp: a pipeline for DNA-seq data analysis. Sci Rep 2018;8 (1):6793.

[58] Gómez-Romero L, Palacios-Flores K, Reyes J, García D, Boege M, Dávila G, et al. Precise detection of de novo single nucleotide variants in human genomes. Proc Natl Acad Sci USA 2018;115 (21):5516−21.

[59] Dewey FE, Murray MF, Overton JD, Habegger L, Leader JB, Fetterolf SN, et al. Distribution and clinical impact of functional variants in 50,726 whole-exome sequences from the DiscovEHR study. Science 2016;354(6319).

[60] Jin Z-B, Wu J, Huang X-F, Feng C-Y, Cai X-B, Mao J-Y, et al. Trio-based exome sequencing arrests de novo mutations in early-onset high myopia. Proc Natl Acad Sci USA 2017;114 (16):4219−24.

[61] Guo X, Shi J, Cai Q, Shu X-O, He J, Wen W, et al. Use of deep whole-genome sequencing data to identify structure risk variants in breast cancer susceptibility genes. Hum Mol Genet 2018;27 (5):853−9.

[62] Khan NS, Saxena P, Pradhan D. De novo sequence assembly. Next Generation Sequencing Data Analysis: integrating genomics, transcriptomics and proteomics data for potential therapeutic target discovery. New Delhi: Biomedical Informatics Centre, ICMR-National Institute of Pathology; 2018. p. 27−8.

[63] Ekblom R, Wolf JBW. A field guide to whole-genome sequencing, assembly and annotation. Evol Appl 2014;7(9):1026−42.

[64] Zerbino DR, Birney E. Velvet: algorithms for de novo short read assembly using de Bruijn graphs. Genome Res 2008;18(5):821−9.

[65] Bankevich A, Nurk S, Antipov D, Gurevich AA, Dvorkin M, Kulikov AS, et al. SPAdes: a new genome assembly algorithm and its applications to single-cell sequencing. J Comput Biol 2012;19(5):455−77.

[66] Luo R, Liu B, Xie Y, Li Z, Huang W, Yuan J, et al. SOAPdenovo2: an empirically improved memory-efficient short-read de novo assembler. Gigascience 2012;1:18.

[67] Jackman SD, Vandervalk BP, Mohamadi H, Chu J, Yeo S, Hammond SA, et al. ABySS 2.0: resource-efficient assembly of large genomes using a Bloom filter. Genome Res 2017;27(5):768−77.

[68] Mikheenko A, Prjibelski A, Saveliev V, Antipov D, Gurevich A. Versatile genome assembly evaluation with QUAST-LG. Bioinformatics 2018; 34(13):i142−50.

[69] Sellera FP, Fernandes MR, Moura Q, Souza TA, Nascimento CL, Cerdeira L, et al. Draft genome sequence of an extensively drug-resistant Pseudomonas aeruginosa isolate belonging to ST644 isolated from a footpad infection in a Magellanic penguin (Spheniscus magellanicus). J Glob Antimicrob Resist 2018;12:88−9.

[70] Thirugnanasambandam R, Inbakandan D, Abraham LS, Kumar C, Sundaram SM, Subashni B, et al. De novo assembly and annotation of the whole genomic analysis of Vibrio campbellii RT-1 strain, from infected shrimp: Litopenaeus vannamei. Microb Pathog 2017;113:372−7.

[71] GATK | Home. [cited 2018 Aug 26]. Available from: https://software.broadinstitute.org/gatk/.

[72] Burrows-Wheeler Aligner. [cited 2018 Aug 26]. Available from: http://bio-bwa.sourceforge.net/.

[73] Pradhan D, Saxena P, Khan NS. RNA-Seq data analysis: differential gene expression analysis. Next Generation Sequencing Data Analysis: integrating genomics, transcriptomics and proteomics data for potential therapeutic

target discovery. New Delhi: Biomedical Informatics Centre, ICMR-National Institute of Pathology; 2018. p. 46–50.

[74] Kim D, Pertea G, Trapnell C, Pimentel H, Kelley R, Salzberg SL. TopHat2: accurate alignment of transcriptomes in the presence of insertions, deletions and gene fusions. Genome Biol 2013;14 (4):R36.

[75] Pertea M, Kim D, Pertea GM, Leek JT, Salzberg SL. Transcript-level expression analysis of RNA-seq experiments with HISAT, StringTie and Ballgown. Nat Protoc 2016;11(9):1650–67.

[76] Ghosh S, Chan C-KK. Analysis of RNA-Seq data using TopHat and Cufflinks. Methods Mol Biol 2016;1374:339–61.

[77] Trapnell C, Roberts A, Goff L, Pertea G, Kim D, Kelley DR, et al. Differential gene and transcript expression analysis of RNA-seq experiments with TopHat and Cufflinks. Nat Protoc 2012;7 (3):562–78.

[78] Love MI, Huber W, Anders S. Moderated estimation of fold change and dispersion for RNA-seq data with DESeq2. Genome Biol 2014;15(12):550.

[79] Robinson MD, McCarthy DJ, Smyth GK. edgeR: a Bioconductor package for differential expression analysis of digital gene expression data. Bioinformatics 2010;26 (1):139–40.

[80] Ritchie ME, Phipson B, Wu D, Hu Y, Law CW, Shi W, et al. limma powers differential expression analyses for RNA-sequencing and microarray studies. Nucl Acids Res 2015;43(7):e47.

[81] Varet H, Brillet-Guéguen L, Coppée J-Y, Dillies M-A. SARTools: a DESeq2- and EdgeR-based R pipeline for comprehensive differential analysis of RNA-Seq data. PLoS One 2016;11(6):e0157022.

[82] Yu G, Wang L-G, Han Y, He Q-Y. clusterProfiler: an R package for comparing biological themes among gene clusters. Omics J Integr Biol 2012;16(5):284–7.

[83] Huang DW, Sherman BT, Lempicki RA. Systematic and integrative analysis of large gene lists using DAVID bioinformatics resources. Nat Protoc 2009;4 (1):44–57.

[84] Cline MS, Smoot M, Cerami E, Kuchinsky A, Landys N, Workman C, et al. Integration of biological networks and gene expression data using Cytoscape. Nat Protoc 2007;2(10):2366–82.

[85] Carlin DE, Demchak B, Pratt D, Sage E, Ideker T. Network propagation in the cytoscape cyberinfrastructure. PLoS Comput Biol 2017;13(10):e1005598.

[86] Haas BJ, Papanicolaou A, Yassour M, Grabherr M, Blood PD, Bowden J, et al. De novo transcript sequence reconstruction from RNA-seq using the Trinity platform for reference generation and analysis. Nat Protoc 2013;8 (8):1494–512.

[87] Kunz M, Löffler-Wirth H, Dannemann M, Willscher E, Doose G, Kelso J, et al. RNA-seq analysis identifies different transcriptomic types and developmental trajectories of primary melanomas. Oncogene 2018.

[88] Arun AS, Tepper CG, Lam KS. Identification of integrin drug targets for 17 solid tumor types. Oncotarget 2018;9(53):30146–62.

[89] HISAT2. [cited 2018 Aug 26]. Available from: https://ccb.jhu. edu/software/hisat2/index.shtml.

[90] Pertea M, Pertea GM, Antonescu CM, Chang T-C, Mendell JT, Salzberg SL. StringTie enables improved reconstruction of a transcriptome from RNA-seq reads. Nat Biotechnol 2015;33 (3):290–5.

[91] SAMtools. [cited 2018 Aug 26]. Available from: http://samtools. sourceforge.net/.

[92] R: the R Project for Statistical Computing. [cited 2018 Aug 26]. Available from: https://www.r-project.org/.

[93] Bioconductor - Home. [cited 2018 Aug 26]. Available from: https://www.bioconductor.org/.

[94] Cytoscape: an Open Source Platform for Complex Network Analysis and Visualization. [cited 2018 Aug 26]. Available from: http://www.cytoscape.org/.

[95] Witzel M, Petersheim D, Fan Y, Bahrami E, Racek T, Rohlfs M, et al. Chromatin-remodeling factor SMARCD2 regulates transcriptional networks controlling differentiation of neutrophil granulocytes. Nat Genet 2017;49(5):742–52.

[96] Guerrero-Martínez JA, Reyes JC. High expression of SMARCA4 or SMARCA2 is frequently associated with an opposite prognosis in cancer. Sci Rep 2018;8(1):2043.

[97] Zhang D, Wang H, He H, Niu H, Li Y. Interferon induced transmembrane protein 3 regulates the growth and invasion of human lung adenocarcinoma. Thorac Cancer 2017;8(4):337–43.

[98] Graña O, López-Fernández H, Fdez-Riverola F, González Pisano D, Glez-Peña D. Bicycle: a bioinformatics pipeline to analyze bisulfite sequencing data. Bioinformatics 2018;34(8):1414–15.

[99] Park S-J, Kim J-H, Yoon B-H, Kim S-Y. A ChIP-Seq data analysis pipeline based on bioconductor packages. Genomics Inform 2017;15(1):11–18.

[100] Afgan E, Baker D, Batut B, van den Beek M, Bouvier D, Cech M, et al. The Galaxy platform for accessible, reproducible and collaborative biomedical analyses: 2018 update. Nucl Acids Res 2018;46(W1):W537–44.

# Chapter | 5 |

# Nuclear magnetic resonance

*Leonardo Vazquez[1] and Gauri Misra[2]*
[1]*Key Laboratory of Pathogenic Microbiology and Immunology, Institute of Microbiology, Chinese Academy of Sciences, Beijing, China,* [2]*Amity Institute of Biotechnology, Amity University, Noida, India*

## 5.1 Introduction

The nuclear magnetic resonance (NMR) technique applied to biological systems is a powerful tool for structural determination and dynamics of both small organic molecules and biopolymers. The dynamic aspects and inherent flexibility play a central role in the facets of biological functionality [1,2]. Regarding natural biopolymers, such as proteins, nucleic acids, and carbohydrates, the structural fluctuations of these molecules can range from nanoseconds to hours, referred to as molecular dynamics [3−5].

The currently available NMR experiments cover a broad spectrum of the observable time-scales phenomena showed by both small molecules and biopolimers (Fig. 5.1). It is possible to monitor structural fluctuations, which may offer information about fast or slow movements. The cellular environment experienced by the molecule includes a wide range of pH, salinity, solutes, and temperature, which exhibits a significant effect on the structure of these molecules [6−10].

The history of NMR starts in the early 1940s, specifically in 1943, with Dr. Otto Stern at the Carnegie Institute of Technology, United States. He received the Nobel Prize in 1944 for the description of the angular momentum of quantized subatomic particles. He proposed electrons and atoms can act as rotational punctual charges and thus, generate a magnetic field on their surroundings. This work, therefore, opened the doors for further studies of the magnetic properties of nuclei [11].

A few years later, Dr. Isidor I. Rabi, Dr. Felix Bloch, and Dr. Edward M. Purcell, all in the United States, established the first effective approaches to study of NMR and data interpretation expanding the need for a new field of study and deepening the comprehension of the composition of different materials at the atomic level [12,13].

Several years later, Dr. Richard R. Ernst (Nobel Prize, 1991) in Switzerland and Dr. Kurt H. Wüthrich (Nobel Prize, 2002), in the United States and Switzerland, contributed to the development of the high-resolution NMR approach for chemistry and the development of an innovative method for mapping and elucidation of the 3D structure of biological macromolecules in solution [14,15].

With the advent of these methodological advances, today it is possible to extract detailed information related to the structure and dynamics of proteins, molecular complexes, carbohydrates, and nucleic acids in atomic resolution [16−19]. Practical NMR applications include:

1. Measurements and description of the molecular dynamics at the atomic level.
2. Modeling of protein and carbohydrate structures. However, there are some limitations regarding the molecular weight of these biopolymers. To overcome this problem, new strategies for data acquisition and differential isotopic labeling are possible options [20−22].
3. Intermolecular interactions are widely used for screening of drugs and ligands in the cell, also including biomimetic membranes [16,23].
4. Understanding the mechanisms behind the protein folding is an important application of this technique, as the structural features, dynamic movements, and interaction measurements collaborate for the precise determination of folding mechanisms [24].

Besides the applications on biopolymers, the advances in the organic chemistry brought precise and productive tools for acquisition and analysis of biological fluids like

**Data Processing Handbook for Complex Biological Data Sources. DOI: https://doi.org/10.1016/B978-0-12-816548-5.00005-8**

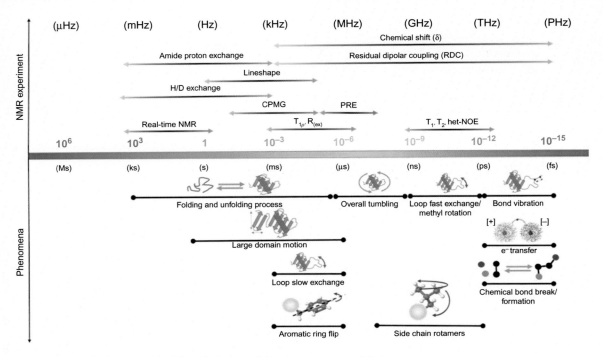

**Figure 5.1** General scheme of the NMR techniques and the phenomena associated.

urine, sweat, and saliva. Furthermore, in the metabolomics, it is possible to find specific metabolites in complex mixtures [25−27]. The structural analysis of organic molecules and natural products were, indeed, greatly benefited from these advances [28−30]. The lessons learned from the experience with small molecules have led to significant applications of these techniques in biopolymers with the development of bi- or multidimensional NMR experiments [31,32].

In addition to these applications, the low field NMR, which presents hydrogen frequencies below 100 MHz, has many areas of application, especially relaxometry experiments (transverse relaxation measurements), hyperpolarization, prepolarization, and diffusiometry assays.

This kind of technique is usually economical compared with the measurements using high field magnets. This technique, allows the researcher to obtain direct data interpretation, besides the commercial acquisition of robust and portable spectrometer, which is highly desirable in industrial applications involving food and crude oil industries [33−35].

In diagnostic medicine, the development of magnetic resonance imaging (MRI) has reached a significant qualitative advance, based on hydrogen nuclei, mainly by the study of two researchers: Dr. Paul C. Lauterbur and Dr.

Peter Mansfield [36]. This technique became the central tool in noninvasive diagnosis and functional monitoring of in vivo systems, helping in the functional evaluation of physiological disorders.

There are some limitations to the use of high field NMR as an analytical method for the description of biological samples in solution. For instance, a protein with high molecular weight may be a challenge on use of this technique, because the dispersion of the magnetization is too fast, which does not allow several pulse sequences [37]. However, in some cases, this limitation can be partially overcome by modifying the physical conditions of the sample such as increase in temperature (it increases the tumbling rate, decreasing the viscosity) and buffer solvent. The problem has to be analyzed for each sample, depending on the purpose of the user.

Moreover, changes in the aggregation state (monomer−dimer−oligomer) should be avoided. Homogeneity and stability are necessary conditions for a suitable NMR sample.

Not only may the size and stability be limiting, but also the sample concentration. It is hard to define a minimum or ideal concentration because this parameter is strongly dependent on the experimental objective and the strategy of the researcher. The NMR is a low sensitive technique and diluted samples may provide futile results. In a high magnetic field ($>18.8$ T, 800 MHz), a collection of meaningful data requires a protein concentration of around 100 μM or even less.

## 5.2 Theoretical and practical aspects

### 5.2.1 Fundamental concepts

In the NMR experiment, the magnetic moment of the atomic nuclei is extensively explored. As a result of the inherent propriety of subatomic particles called spin, which is estimated by approximating the nuclear rotation movement (Fig. 5.2). The spin, in turn, is a mechanical

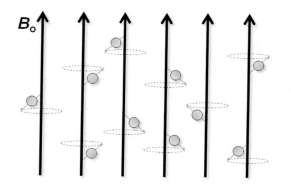

**Figure 5.2** Nuclear precession and alignment of nuclei, when exposed to a strong external magnetic field ($B_0$).

property of subatomic particles that generates the angular momentum component, referred to as the nuclear magnetic moment $\mu$, which has no equivalent in classical mechanics [38,39].

The nuclear magnetic moment $\mu$ of a nucleus is proportional to the Planck's constant ($\hbar$), the gyromagnetic relation $\gamma$ (rad s$^{-1}$ T$^{-1}$), and its spin quantum number $I$, which are specific mechanical properties of each nucleus described in Eq. 5.1.

$$\mu = \gamma I \hbar \tag{5.1}$$

For NMR measurements, to identify magnetically active atomic nuclei, the nuclei should have its quantum number of spin $l$ different from 0 and is defined by the relation $2I + 1$ possible spin states. The magnetic quantum number $m$ varies as follows: $-I, I+1, I-1, I$. Therefore, according to this relation, the nuclei with $I = \frac{1}{2}$ can present only two possible spin states: $m = +\frac{1}{2}$ ($\alpha$) and $m = -\frac{1}{2}$ ($\beta$).

The most common nuclei used in experiments with biological samples include the following isotopes: $^1$H, $^{13}$C, $^{15}$N, $^{19}$F, and $^{31}$P [40]. However, the natural abundance of two of the major constituent of organic biomolecules are $^{13}$C and $^{15}$N, having values of approximately 1.108% and 0.37%, respectively [41]. Biological samples are known to be a challenge to reach enough concentrations detectable by NMR. Thus, the choice is the isotopic enrichment using the commercial bacterial host *Escherichia coli* strains. The source of desired isotopes is U-$^{13}$C-glucose ($^{13}$C$_6$H$_{12}$O$_6$), for $^{13}$C labeling, and U-$^{15}$N-ammonium chloride ($^{15}$NH$_4$Cl) for $^{15}$N labeling [42,43]. Table 5.1 shows specific constants of the essential nuclei used in the NMR of biomolecules.

Analyzing the table, the nucleus of $^1$H is ubiquitous in biomolecules ($>99.9\%$ natural abundance) and presents the higher gyromagnetic relation $\gamma$. It results in higher

**Table 5.1 Physical properties of most common nuclei used in NMR experiments.**

| Nuclei | Nuclear spin | Natural abundance (%) | $\gamma$ ($10^7$ s$^{-1}$ T$^{-1}$) | Larmor frequency (MHz) ($B_0 = 2.35$ T) | Larmor frequency (MHz) ($B_0 = 9.4$ T) | Larmor frequency (MHz) ($B_0 = 11.74$ T) |
|---|---|---|---|---|---|---|
| $^1$H | 1/2 | 99.985 | 26.752 | 100 | 400 | 500 |
| $^2$H | 1 | 0.015 | 4.107 | 15.35 | 61.40 | 76.75 |
| $^{13}$C | 1/2 | 1.108 | 6.728 | 25.14 | 100.56 | 125.72 |
| $^{15}$N | 1/2 | 0.370 | $-2.712$ | 10.13 | 40.52 | 50.66 |
| $^{19}$F | 1/2 | 100.000 | 25.181 | 94.08 | 376.32 | 470.39 |
| $^{31}$P | 1/2 | 100.000 | 10.841 | 40.48 | 161.92 | 202.40 |

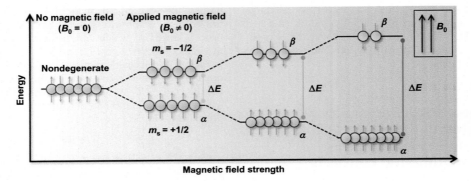

**Figure 5.3** Zeeman effect or split caused by the application of a strong magnetic field to nuclei population.

sensitivity rendering this nucleus a preferred choice for both spectral acquisition and magnetization transfer mechanisms in multidimensional experiments.

At the start of the NMR experiment, the nuclear spins states are degenerate, or their energetic quantum states are located in equivalent levels. The basis of this technique is the result of defined quantum energetic states experienced by a population of nuclei. A strong external magnetic field ($B_0$) applied to a set of degenerate nuclei disturbs the symmetry and gives rise to the split into two distinct levels ($\alpha$, $\beta$).

The $\alpha$ state of lower energy is the fraction of spins that aligns in the same direction as the external field $B_0$. On the other hand, the $\beta$ state, with higher energy, refers to the fraction of spins aligned against the direction of external field $B_0$. This split in the spin states is termed the Zeeman effect or separation (Fig. 5.3). The energy associated with each adopted state ($E_m$) is dependent on the intensity of the external magnetic field $B_0$ and the value of the gyromagnetic relation $\gamma$, see Eq. (5.2).

$$E_m = -m\gamma\hbar B_0 \qquad (5.2)$$

The difference of energy ($\Delta E$) between two nuclear spin states is the function of the external magnetic field and the gyromagnetic constant of this nucleus, as described by the Zeeman Eq. (5.3).

$$\Delta E = \gamma\hbar B_0 \qquad (5.3)$$

An essential concept in NMR is the Larmor frequency $\omega_0$ (rad s$^{-1}$), which is the value of the frequency of a given nucleus explained in the Eqs. (5.4) and (5.5):

$$E = \hbar\omega_0 = \gamma\hbar B_0 \qquad (5.4)$$

We have:

$$\omega_0 = \gamma B_0 \qquad (5.5)$$

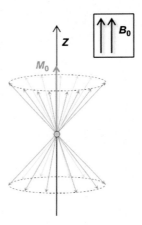

**Figure 5.4** Schematic representation of the macroscopic magnetization caused by unequal distribution of a set of individual nuclei spins.

The number of nuclei thermodynamically positioned in the lower energy state $\alpha$ is only slightly higher than the nuclei, which adopts the higher energy level $\beta$. The description for this property is based on the statistics of Boltzmann distribution or Maxwell–Boltzmann distribution and results in the relation below, where $N_\alpha$ and $N_\beta$ are the fractions of the total nuclei in each state, $\Delta E$ is the difference of energy of both states.

$$\frac{N_\beta}{N_\alpha} = e^{-\frac{\Delta E}{kT}} \qquad (5.6)$$

External factors modify this split, including the intensity of the magnetic field, understood by the influence on the factor $\Delta E$ and the absolute temperature of the system ($T$).

The slight excess of nuclear spins in the same direction as the magnetic field $B_0$ is termed as macroscopic or liquid magnetization (Fig. 5.4). Since, intrinsically, the population difference between states $\alpha$ and $\beta$ is low, the resulting vector of measurable magnetization is also low, rendering NMR to be a low sensitive technique [44].

The resulting spectrum is a collection of scans when summed causes accumulation of signals at specific frequencies that represents a real resonance signal, not a random noise signal. So, an important parameter is the signal-to-noise ratio (SNR) given by the Eq. (5.7), where $N$ represents a number of scans.

$$SNR \propto \sqrt{N} \tag{5.7}$$

Some notes regarding the signal-to-noise ratio follow:

Consider the phase cycling necessary to calculate the number of scans (check the pulse program file of the experiment), for example, phase cycling usually occurs in multiples of 4 steps, so the number of scans will then follow multiples of 4 [4,8,16,32]. The noise signal is random, so the intensity does not increase during the acquisition.

The NMR measurement represents the change in the modulus and the direction of the macroscopic magnetization vectors previously aligned in the same direction of the $B_0$ field (in equilibrium from the z-axis). At rest state, the z component of the magnetization ($M_z$) has the same modulus as of $M_0$.

## 5.2.2 Nuclear magnetic resonance relaxation mechanisms

The time required for an excited state of the spin to return to the fundamental state, along the z-axis (+z), is governed by the spin-lattice relaxation $T1$, the following Eq. (5.8) describes this behavior.

$$M_z = M_0(1 - e^{-\frac{t}{T1}}) \tag{5.8}$$

Considering the longitudinal magnetization in the ($-z$) direction. Eq. (5.9) describes, as a function of time, the relaxation along the z-axis back to the initial rest state ($M_0 = M_z$)

$$M_z = M_0(1 - 2e^{-\frac{t}{T1}}) \tag{5.9}$$

When a radio frequency pulse is applied on the nuclei population in the $\alpha/\beta$ states, it changes to the excited state of higher energy ($\beta$) causing the change in the macroscopic magnetization (Fig. 5.5).

The effective magnetic field applied to each nucleus after the excitation is slightly different, depending if the nucleus is bound covalently or not and its positioning nearby to another nucleus. These interactions are namely spin–spin interactions. In this situation, the Larmor frequency will be slightly altered and represents another source of relaxation. Thus, the chemical neighborhood is the primary influence on the loss of coherence of the spins after excitation in the XY plane ($M_z \sim 0$). The transverse relaxation occurring in the XY plane (Fig. 5.6) is governed by Eq. (5.10), where the $M_{xy}$ is the value of the modulus of the magnetization on the transverse plane.

$$M_{xy} = M_{xy_0}e^{-\frac{t}{T2}} \tag{5.10}$$

Note: In general, $T2$ values will be lower or equal to $T1$ [45]. These two magnetization phenomena occur at the same time, during the nuclear relaxation. The transverse relaxation goes to zero (loss of coherence) at the same time as the magnetization vector on the z-axis return to the initial position. Despite this behavior, they are independent; one process of relaxation does not affect the other [45].

The result of these relaxation phenomena can then be understood as magnetization vectors that oscillate in the distinct axes (X and Y). The signal reception in the magnet is given by a coil positioned along the x-axis, while the magnetization is located in the xy plane. Thus, the recording of the electric current generated in the coil caused by the variation of the magnetization in the xy plane, as a

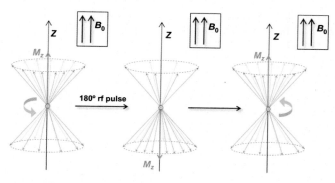

Figure 5.5 Schematic representation of the longitudinal relaxation (T1).

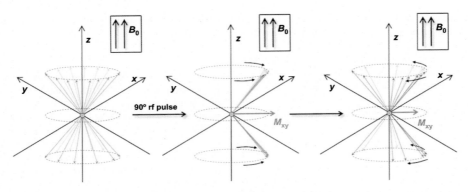

**Figure 5.6** Schematic representation of the transversal relaxation (T2), on the last scheme, the hot colors shows the faster dephasing spins, while the cold colors depict the slowest dephasing spins.

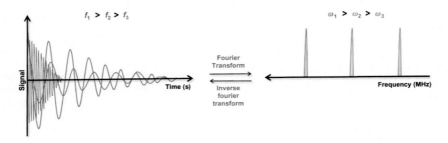

**Figure 5.7** Scheme of the Fourier transform, which transforms the plot initially on the time domain to the frequency domain and vice versa. Frequencies representations out of scale.

function of time, generates a sinusoidal signal called FID (free induction decay, Fig. 5.7). The dephasing of spins causes the natural decay in amplitude in due course of time during the spin—spin relaxation. For a better interpretation of the raw data provided by the magnet, the mathematical operation called Fourier transform is used. This formula is applied to the functions in the time domain, which in turn will be decomposed into their sine and cosine components in the frequency domain. Also, the inverse Fourier transform (IFT) converts functions from the frequency domain to the time domain (Fig. 5.7).

As previously discussed, the Larmor frequency is a function of the external magnetic field. Thus, it is complicated to compare two spectra obtained from magnets with different field intensities, resulting in different frequency values. To reconcile these values, the ppm notation is used ($\delta$ or $\omega$, parts per million), so that, based on the internal reference ($\delta = 0$ ppm), the chemical shift is calculated allowing the comparison of different spectra using the following relationship:

$$\delta, ppm = \frac{\text{Frequency of a given nucleus (Hz)} - \text{frequency of the internal standard (Hz)}}{\text{Spectrometer frequency (MHz)}}$$

Note that the frequency is given in the numerator in hertz, and in the denominator in megahertz, so the result of this quotient will be in parts per million (ppm). The usual internal standard is tetramethylsilane (TMS), 0.03% (v/v) that serves as a reference for the acquisition of $^1$H, $^{13}$C and $^{29}$Si spectra, since it is soluble in organic solvents and can be easily removed by its low boiling point (26.6°C). It shows almost no interaction with samples in general and presents only a single reference signal in the spectrum ($\delta = 0$). In NMR of biological molecules, where water is the solvent, it is possible to use sodium 3-trimethylsilyl propane sulfonate (DSS) in deuterated water [46].

Representation of 1D experimental spectrum involves the chemical shift unit in parts per million $\delta$ (ppm) plotted on the x-axis, while the y-axis corresponds to the intensity in arbitrary units [45]. In 2D or multidimensional experiments, both axis are identified in chemical

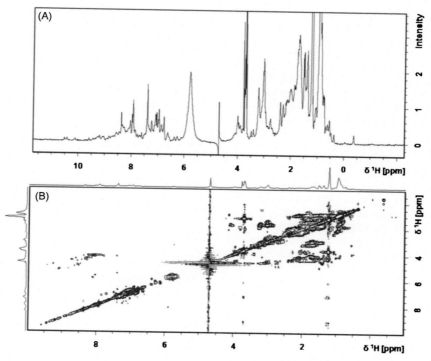

**Figure 5.8** NMR spectra of protein: (A) 1D $^1$H-NMR spectrum is an example of well-folded protein, which shows typical characteristic peaks below 1 ppm and well-dispersed peaks on the amide region (around 8 ppm). (B) [$^1$H, $^1$H] TOCSY spectrum shows the correlation between hydrogen atoms up to three covalent bonds far. The presence of diagonal correlation is characteristic of the homonuclear spectrum.

shift ($\delta$ [ppm]) of the corresponding nucleus, but the intensity is plotted in the contour plot (the number of contours can be defined in the acquisition software). Fig. 5.8 shows an example of such spectra.

The origin of the chemical shift is multifactorial because it includes characteristics of the molecule itself and its environment. The characteristics to be considered are the concentration of the sample, pH, solvent composition and the temperature since these factors significantly alter the behavior of the biological polymers in solution. As discussed previously, the amount of information obtained from NMR is considerable, so in a detailed analysis of the spectra, any change in these parameters can compromise the final results obtained from the spectra [44,47]. The anisotropy of molecules and orbitals, intermolecular interactions as hydrogen bonding, salt bridges, hydrophobic interactions, paramagnetic ions in solution or interactions (some transition metals, and more significantly the lanthanide series) are additional factors that will alter the chemical shift [18]. The defined force of magnetic interaction of two nuclei is referred to as coupling constant.

Scalar coupling or spin—spin splitting is the interaction between covalently bound atoms represented by the notation: $J_{AX}$, where A, X represents any pair of atoms.

The nuclear polarization of atom A disturbs the electrons of the bond (A—X) and then polarizes the bonding atom X. This value is essential since it is always constant for a specific magnetic interaction between two covalent bonded atoms (independent of the external magnetic field in ppm value). This characteristic is also useful to recognize the defined spatial arrangement of atoms in a molecule and used to help in the structural determination by NMR [45]. The notations $^1J$, $^2J$, and $^3J$ represent the distances of one, two, or three respectively covalent bonds far from any active magnetic atom in a molecule. Alternatively, the influence of couplings longer than these as $^4J$, $^5J$ is not observable in the traditional experiments [45].

Interactions that do not involve magnetic interactions mediated by electrons in chemical bonds but only by spatial proximity are known as dipolar interactions or dipole—dipole coupling. The notation is $D_{AX}$ or $d_{AX}$, where A, X represents any pair of atoms. On the other hand, the use of cross-relaxation of nuclear spins through the nuclear effect overhauser (NOE) is the main approach for the spatial elucidation of proteins and carbohydrate structures or organic molecules, estimating distances up to 5 Å between a pair of nuclei [45].

Still, in this context, it is essential to emphasize the role of solid-state NMR (ssNMR). This technique allows the researcher to work with aggregates, high viscosity environments, vesicles, or any other low diffusion context. The main characteristic of this technique is to overcome the strong anisotropic interactions between molecule with almost no Brownian motion and the dipolar interactions present in these molecules, which enhances the rate of relaxation broadening the line shape in the spectrum. In order to obtain better resolution, the sample is spinned from 1 to 130 Hz, additionally, is tilted with a precise angle of 54.74 degrees, the magic angle. These procedures aim to cancel the strong dipolar coupling and enhance the mobility under high viscosity conditions [48].

The result is better resolution with less broad peaks, which otherwise would be impossible to analyze since the width of the peaks and overlap will disturbing the identification of individual resonances. It is crucial to mention that the resolution in NMR refers to the possibility of distinguishing clearly between two distinct peaks, in other words, the separation of lines in the spectrum [49,50].

The sensitivity refers to the distinction between the background noise and a real resonance signal. Often these two characteristics may be associated because when there is a loss of resolution (peak broadening) there is associated decrease in sensitivity (peak height).

In NMR measurements, the electron shielding is the primary effect observable that happens by the movement of the electrons in the orbit of nuclei. In this way, the value of the field effectively applied to a given nucleus will be different of $B_0$. The electron shielding shifts the position line in the spectrum, often to the right side, strengthening the effect. The high field of the spectrum, the values of the chemical shift are smaller, usually <3 ppm, when considering hydrogen frequency. At the same time, the value of frequency will also be lower. On the other hand, when the electron density is lower, the strength of the field $B_0$ will be more effective. In this case, the nucleus is deshielded. Under such conditions, it is observed that higher values of the resonance frequency, higher values of chemical shifts in the low field region are usually located to the left side of the spectrum [51].

The inductive effect may be the simplest way to understand the mechanism of this electron density displacement [45]. The stable state of the partial polarization of the bond between any two atoms represents this change in the local electron density. The force of the inductive effect is a factor of the distance between the two nuclei covalently bound and the difference in the value of electronegativity between them. It is also important to note that the electron density is also related to the presence of $\pi$ resonance (double and triple bonds) and the bond geometry. In all cases, the practical result of this phenomenon is the asymmetric distribution of charges in the polarized bond. When the electronegative atom is the primary source of asymmetry in the bond (electron attractor), it is termed as negative induction [52].

Single and multidimensional resonance experiments such as rotating-frame Overhauser effect spectroscopy (ROESY) and nuclear Overhauser effect spectroscopy (NOESY) [53,54] allow for an integrated structural data analysis involving chemical bond torsion, and distances between unrelated nuclei defined by chemical bonds. These experiments are useful for determining the signals that arise from protons related through space or covalently bonding [55].

## 5.3  Raw data

NMR experiment starts with proper data collection and magnet adjustment. The quality of the FID data depends significantly on the sample conditions and a proper equipment adjustment. A handy guide to the procedure followed for initial NMR data acquisition of a quality spectrum is briefly outlined below:

1. Check the reference data that was recorded at the time of the instrument installation to verify that the specifications of the manufacturer are satisfied. The most important data are:
   1.1 Sensitivity test with ethyl benzene (0.1% in $CDCl_3$), Fig. 5.9A.
   1.2 Sensitivity test with an aqueous sucrose solution.
   1.3 "Hump test" with $CHCl_3$ (1% in Acetone-$d_6$) sample, Fig. 5.9B.
   1.4 90 degrees pulse length for $^{1}H$, $^{13}C$, and $^{15}N$.
   1.5 Measurement of the probe temperature, using the reference samples provided by the manufacturer. Compare the measured temperatures with the temperature dial on the spectrometer console.
2. Perform the measurements 1.1−1.5 with the standard samples provided by the manufacturer, which must be available in the NMR laboratory. Compare the results obtained with the reference data. To obtain results close to or equal to the reference data, shimming of the magnetic field and tuning of the probe will be required.
3. With the instrument tuned and shimmed for optimal performance with the aqueous sucrose solution, record first 1D $^{1}H$-NMR spectrum with a protein

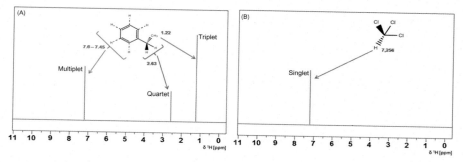

**Figure 5.9** Representation of 1D $^1$H-NMR spectra of standards. (A) 1D $^1$H-NMR spectrum of the ethyl benzene, a chemical used to test the sensibility of the probe. (B) The 1D $^1$H-NMR spectrum of the $CHCl_3$ is used for the "hump test," where the symmetry, flatted baseline, and Lorentzian shaped peak are the aim.

solution and check again, if tuning and shimming are optimized for working with the sample.

4. Record the 2D ($^{15}$N, $^1$H)-HSQC experiment as a reference, documenting the quality of the protein sample at the start of your experiment.
5. Perform the long-time experiment.
6. Record the 2D [$^{15}$N, $^1$H]-HSQC experiment to check by comparing with the experiment 4, ensuring that the protein solution was not altered during the long-time experiment.

If the recording data will be performed using a familiar instrument, the protocol will be following:

1. Record the tests 1.2, 1.4, and 1.5. Do not continue if the tests give poor results (e.g., low sensitivity, poor line shape, long pulse lengths). Check probe tuning and instrument shimming. If you do not arrive close to the reference, request help from the technical specialist.
2. Only if A. provides satisfactory results, proceed with the steps 3−6 above.

The requirements for the biological samples include adequate sample tubes, usually filled with the volume of 500 μL. An estimated concentration of 1 mM of the sample (DNA/RNA or protein) is desirable, but not mandatory, for structural elucidation. Concentrations around 150 μM in a powerful magnetic field and altered physical conditions as temperature, chemical additives, and pH are sufficient to obtain an adequate spectrum. Chemical impurities may compromise the analysis of the data, once that the signals related to the contaminants will appear overlapped with the spectra of the sample.

A reference buffer used in NMR samples consists of 10−50 mM phosphate buffer pH 7.0, 50−100 mM NaCl and optionally 50 mM EDTA and 3 mM of azide. The buffer composition may be altered depending on the

experimental condition. If the requirement of work is extreme pH values then it is recommended to use buffers with poor ionic mobility [56]. The presence of an internal standard is mandatory together with the sample. Usually, the solvent composed of 10% (v/v) $D_2O$ in 90% $H_2O$ is the choice for biological samples. When the aim is to measure exchangeable protons, the biological sample is lyophilized and then redissolved in 100% $D_2O$, at low temperature and pH, as the exchange rate will be slower [57,58].

The raw NMR data is the unprocessed FID. Moreover, it has time-domain intensity values (seconds) that will then be processed for the frequency domain in Hz and calculated for ppm using the Fourier transform [44]. The raw data should be processed after the acquisition of the complete scans. For 1D data, the output file name is *fid*. The plotted result is the signal from the equipment in the time domain (Fig. 5.10A) for 2D and 3D data, the file is *ser* (Fig. 5.10B and C, respectively). However, the researcher can find this file in the folder of the data set collected. Once this file is moved, the program will no longer recognize the data set and further processing will not be possible.

For suitable data acquisition, the following steps are involved:

1. Check the nitrogen and helium levels of the equipment:
   This procedure avoids inconveniences of temperature variation or even quenching of the magnetic field.
2. Check the temperature on the probe:
   When the sample is inserted in the instrument (*lift off*), it is possible to follow the evolution of the temperature stabilization, since the field settings are highly dependent on the temperature variation in the probe.

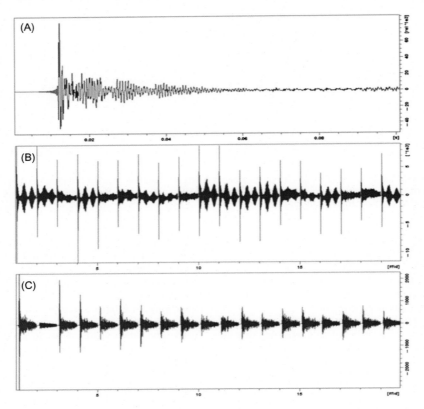

**Figure 5.10** Raw data: (A) Simple FID from 1D ¹H-NMR experiment. (B and C) A serial set of FID from 2D and 3D experiments, when the set of FID builds the indirect dimension, usually nitrogen or carbon in protein and nucleic acids.

**3.** Attention to "lock" signal:

For biological samples, where water is the primary solvent, the deuterated water is used as a reference to stabilize the strength of the magnetic field showing the signal on the lock display of the equipment. The origin of the lock signal occurs through the constant pulse of the deuterium channel in the sample. Thus, it allows the operator to adjust an effective field, phase, and power, as well as it helps to monitor the strength of the pulse applied to the sample. Before the start of the first adjustment it is mandatory that the operator chooses the correct lock system based on the composition of the solvent.

**4.** Stability of temperature: once that the probe temperature is stable, observe the parameters of power and phase of the standard pulse. These parameters must be adjusted so that the signal is symmetrical (phase) with the lowest noise (power). The company provides several standards which are adequate for this purpose. However, a user's sample is the most

common way to adjust these parameters, but attention is required to the presence of high salt concentration. This condition significantly modifies the pulses in the deuterium channel, rendering ineffective lock status, usually recovered by manual adjustment.

**5.** Tuning and matching:

The sensitivity of any probe will vary according to the correct adjustment of tuning and matching. The tuning parameter can be monitored on the screen during the process. The objective is to reach the frequency as close as possible to the nucleus in question; the movement of the curve occurs laterally. In the matching parameter, the ideal position of the curve is that it reaches the lowest possible value, as close as possible to the baseline. Consider repeating the protocol for all the nuclei in the experiment (usually first ¹H, then ¹⁵N or ¹³C).

**6.** Adjustment of shimming:

The shimming adjustment aims to homogenize the strong magnetic field of the magnet using a set of external coils to produce small magnetic fields and removing imperfections on the main magnetic field. The advantage of this procedure is to obtain the narrowest line as possible, which will be decisive on the resolution and sensitivity on the final spectrum. In practice, as standard, the $^1$H line width should be less than 0.5 Hz, but the best line shape as possible and even better sharpness is the main target. On the modern magnets, the deuterium channel is often used to follow the progress adjustment on the NMR lock screen. When fine adjustments (high order shims) is necessary, a $^1$H NMR signal is used. In terms of biological samples, the chemical environment has a decisive implication on the final quality of the shimming process such as: high salt concentrations ($>500$ mM), aggregates or other solid material in suspension, pH fluctuation, temperature modification, and unusual aqueous/solvent mixtures, paramagnetic metals, sample viscosity, sample concentration, mismatch of the tube on the spinner and the coil, and the height of the sample column on the tube.

The collection of a simple 1D $^1$H-NMR experiment is essential to qualitatively evaluate the shimming adjustment routine. A desirable shimming will result in a 1D $^1$H-NMR spectrum with both symmetrical and sharp peak, absence of any kind of shoulder, double tips, and distortions.

7. Pulse calculation:

The level of the 90 degrees pulse can be calculated automatically by the software, which apply a set of template pulses and calculation. These procedures determine an optimized value. This pulse is the basis of calculation for adjusting the power of the secondary pulses and gradients during the pulse sequence, so it is crucial that the user calculates and applies this procedure before the preacquisition step.

8. Adjustment of acquisition parameters:

The acquisition step allows the researcher to set the experimental parameters such as pulse sequence, number of points, spectral size, receiver gain, dummy scans, and number of scans. Special attention is dedicated to transmitter frequency offset. The saturation pulse in most experiments has the offset as a reference, usually centered on the frequency of the hydrogen present in water, namely "O1," and has, in ppm, the value around 7.4. For better accuracy, the adjustment of this parameter is made manually, using an interactive acquisition mode.

9. Finally, with all requirements adjusted, the user can then start the experiment using the command 'zg'.

Following these steps, the user will perform a quality NMR spectrum and ensure the efficient extraction of information from the sample.

## 5.4 Data processing

Once the raw data represented by the FID is obtained, the next step is to perform data processing. Software such as TopSpin [59], Spin works [60], CCPN [61] works on similar lines to find the file fid and process the data. A tutorial for fast and primary processing of 1D, 2D, and 3D spectra is shown further. The data bank of the magnet has its own more than 1500 different NMR experiments that can be processed similarly. For more details, check TopSpin, NMR Data Publishing, User Manual. Basic steps used for manual processing are:

1. Export and open the directory to the processing software (TopSpin [59], Spin works [60], CCPN [61], etc.). Ensure that the fid file exists in the correct folder.

2. Add the autoprocessing commands. Software like Spin works [60], the command of processing is interactive and is possible to handle the data set interactively, on the other hand, the TopSpin [59] software needs the command line and the option for automatic processing is available in the menu.

Consider observing the proper characteristics of the experiment such as water suppression signal, saturation level, undesirable clipping effects ("sinusoidal aspect" surrounded by the most intense signals in the processed spectrum) and unusual defects in the spectrum.

3. Baseline and phase correction. Usually, these adjustments are dealt both automatically or manually. Automatically the baseline is corrected using the command "abs" for 1D spectra, and "abs2" and "abs1," in this order, is applied in 2D spectra and works the major of acquisitions. The phase correction is also available in the processing menu TopSpin software [59] where the inversion of the spectra rotating 180 degrees, usually resolve many problems. The auto phasing correction is applied with the following the command: "apk," "apk0," "apkm," and "apks," only for 1D processed spectra (command: "pro"), for 2D processed spectra (command: "xfb") the commands should be phc0 and phc1, in the new window, the values (in degrees) must be filled. Opening the editor using "edp" command is an alternative way to fill those values. On the other hand, manually, the phase correction should be made using a reference region (pivot) usually is used one isolated peak in the distant region of the spectrum, which will be the reference for the zero order correction or frequency independent (ph0). The correction will be more significant around the pivot point. The first-order correction or frequency dependent (ph1) [1],

used to phase another peak distant from the zero order referenced earlier.

A flat and regular baseline is the objective across the whole spectrum. As mentioned, the choice of pivot position is made using an isolated region or peak of the spectrum with a relative few peaks around and well visible for better evaluation of the modifications applied to the spectrum. For two-dimensional spectrum, the manual phasing is an alternative; the procedure is the same as described previously.

The spectrum can be processed several times, once the FID of the experiment is available, but a wrong collection of data cannot be improved or repaired later. Once, collected in a bad shape, consider revising the acquisition parameters and repeat the experiment. A rigorous review of the parameters set on the acquisition data section is determinant to avoid unexpected useless data.

## 5.4.1 Integrating data approaches including web-based sources

The online tools are essential both before and after the acquisition since they provide useful strategies for improvement in data analysis. Furthermore, the following session deals with some of the essential software used for data analysis. The database sources are updated continuously and work as a valuable tool to help in data interpretation. It is also desirable that the user should be familiar with some basic knowledge in Linux and computing in general, since some of the updates or processing procedures may require operations on these platforms, as well as to extend the experience and update of the software through the forums. Such resources have been summarized in Table 5.2.

**Table 5.2 List of few web-based sources useful for NMR experiments.**

| S. no. | Resource name | Application(s) |
|---|---|---|
| 1. | Biological Magnetic Resonance Data Bank, BMRB | 1. Repository of NMR spectroscopy data for biological molecules such as proteins, peptides, nucleic acids. <br> 2. Used for data validation, metabolomics, diverse calculators, and online data integrating approaches. |
| 2. | Protein Data Bank, PDB | 1. Repository for three-dimensional structures of nucleic acids, proteins and molecular complexes. <br> 2. Provides services for structure validation, structural analysis, alignment and structure quality. |
| 3. | Bioinformatics Resource Portal, ExPASy | 1. Integrated tools for nucleic acids, protein and biological molecules data analysis and prediction. <br> 2. Tools include alignment of DNA sequences, RNA and proteins, identification of small sequences, structure prediction, sequence translation, protein structure homology, and identification of posttranslational modifications. |
| 4. | Universal Protein Resource, UniProt | 1. Focused on protein annotation, sequence retrieval, and functional information. <br> 2. Biological functionalities, sequence conservation, taxonomic relationships, subcellular localization and correlation with other databases can be determined, complementary tools such as UniProt KB also present for functional information of proteins based on primary sequence identity, taxonomic description, and data from literature. <br> 3. The other available tool is UniRef, which works by analyzing clustered sequence sets from UniProt databases. |
| 5. | Basic Local Alignment Search Tool, BLAST | 1. Proteins and nucleic acid sequence analysis. <br> 2. Useful in evolution studies, identification of proteins, genes and structural genomics. |
| 6. | Public Chemistry, PubChem | 1. Contains small molecules along with some larger molecules spectra. <br> 2. Structure of carbohydrates can be analyzed and deposited. <br> 3. Database consists of drugs, carbohydrates, lipids, chemically modified organic molecules, peptides, and nucleotides. |
| 7. | Glycan Repository, GlyTouCan | International repository of glycans. |
| 8. | Spectral Database for Organic compounds, SDBS | Focused on organic compounds and data NMR spectra related to small natural molecules. |

## 5.5 Essential analysis and processing software

### 5.5.1 Acquisition and initial processing

The largest company of NMR magnets currently offers software for the user to process their data on a personal basis. The pulse programs, phase processing, pulse time calculations, and intensity are all available in this software. It is used to extract figures, integrate data, and process spectra for use by other processing platforms. Both the output and the input data of this software adopt the standard Bruker model.

The output names of the processed spectra are "1r" (1D), "2rr" (2D), and "3rrr" (3D). These files will be exported to processing and resonance attribution software. In addition, the user must read the proc (proc), acquisition (acqu), and pulse sequence (pulse program) files of the experiment that will provide valuable description of the data and illustrate information about the scheme of acquisition, which is important for further analysis. It is available for Linux, Windows, and Mac OS X operating systems on the official website: https://www.bruker.com/service/support-upgrades/software-downloads/nmr/free-topspin-processing/free-topspin-download.html

### 5.5.2 Computer-aided resonance assignment: spectra assignment

The CARA software [62] was created for the sole purpose of processing NMR spectra of biological molecules by Prof. Kurt Wüthrich research group. The software enables the user to work with the attribution of backbone resonance proteins, side chains, and tools for spectral marking useful in structure determination (NOESY). A project is created initially, the user so, should have to include all the NMR experiments recorded. After that, identify carefully all experiments added on the project. Start the pick peaking process based on the characteristic chemical shifts of the related residue (e.g., assign the resonance of all $\alpha$ and $\beta$ hydrogens in the HBHA(CO)NH experiment). Once, all the spectra assigned, the final lists will be used for integration on the modeling software.

The software uses XML format to organize the NMR data. The software still performs cross-validation of assignment to validate the marked peaks and inconsistent assignments. The software is compatible with Bruker formats and is available for Windows 9x/2k/XP/Vista/7/8/8.1/10, Linux x86, Mac OS X x86, Mac X11 (//cara.nmr.ch/doku.php/cara_downloads).

### 5.5.3 Spin works (Manitoba University): spectra processing and analysis

The Spin Works software [60] is an open source software for processing one-dimensional and two-dimensional data from NMR. For the analysis of organic molecules, organic natural products, and inorganic chemistry this software is a valuable tool.

An accurate process of data is available. Since the FID data, up to the final processing the tools available are intuitive and helps the user to learn and obtain a well processing set of data. For more information, protocols, and updates visit the developer's web site. (Dr. Kirk Marat, Univ. of Manitoba) at http://home.cc.umanitoba.ca/~wolowiec/spinworks/index.htmlthatis available for platforms Windows 95, 98, NT 4.0, Windows 2000 Pro, or XP (NT 4.0, XP or Win 2000 recommended) and also supports Bruker and Variant outputs.

### 5.5.4 NMRFAM-SPARKY: assignment of integrated spectra

The NMR Facility at Madison (NMRFAM) is currently in charge of updating and distributing this software available with manual and tutorials at https://nmrfam.wisc.edu/nmrfam-sparky-distribution/. This software is intended for NMR analysis and assignment of biomolecules such as polymers, nucleic acids, proteins, and peptides [63].

The input of the program is the previously processed spectrum. The output data is a text file with a list of assigned peaks having specific resonance and the spectral data of volume and line width. The structural determination output files are compatible for most of the modeling software, but a file conversion of extensions is necessary.

Another essential tool is the automatic pick peak that SPARKY can provide but it is strongly suggested to critically review the automatic picking to avoid any inconsistent picking. It is indicated for beginners with the intent to calculate the structure of biopolymers in general. The software is available with its manual at http://www.cgl.ucsf.edu/home/sparky/.

### 5.5.5 Collaborative computing project for nuclear magnetic resonance: spectral processing and analysis

This software is an open source NMR software suite. An international committee has made this increasingly efficient software available for both beginners and advanced users. This society encourage the members promoting international meetings, courses in order to disseminate computational development tools and fruitful discussions [64].

There is a specific advantage of this software because the constant review, updates and feedback from the users increase the efficiency of the tool. The website is complete and is available for download at https://www.ccpn.ac.uk/. Among the utilities presented in this program we can mention automatic pick picking, integration and parallel processing of spectra, extraction of lists for relaxation analyzes and protocols and acquisition methods for solid-state NMR, practical guides for NMR acquisition and more, available for Windows, Linux, Mac OS X, and Unix.

### 5.5.6 Nuclear magnetic resonance pipe processing and analysis

The NMR pipe [65] is a comprehensive program that is geared towards data processing from the NMR magnet, available for download at https://www.ibbr.umd.edu/nmrpipe/install.html. This program includes nonuniformly sampled (NUS) method processing, which works using a random data acquisition and significantly reduces data acquisition times and can increase the spectral resolution increasing the number of sampling points [66]. Experiments with more extensive sampling require longer acquisition time, where there is considerable relation between better resolution and bandwidth. This software requires effective UNIX knowledge and scripts. Compatible with UNIX, Linux, and Mac OS X.

## 5.6 Structure determination softwares

See Table 5.3.

**Table 5.3 List of softwares used for structure determination and analysis.**

| S. no. | Resource name | Application(s) |
|---|---|---|
| 1. | ARIA [67] | 1. Automated NOE assignment and structure calculation.<br>2. Compatible with other software as CCPN. |
| 2. | CYANA [68] | 1. Automatic NOE assignment and structure calculation.<br>2. Less processing requirement, useful and more efficient than other methods. |
| 3. | UNIO/J-UNIO [69] | 1. Automated structure calculation, combined with the intuitive graphical interface.<br>2. Different algorithms for assignment as MATCH algorithm [70], for protein side chains. The ASCAN algorithm [71], the automated NOE already assignment by the CANDID algorithm [72], and the ATNOS algorithm for NOESY peak picking [73].<br>3. Includes the basic peak picking for processed APSY (Automated Projection Spectroscopy) experiments.<br>4. The complete assignment process include the analysis of NOESY experiments: 3D Ali$^{13}$C {$^1$H, $^1$H}-NOESY, 3D $^{15}$N-{$^1$H, $^1$H}-NOESY, 3D $^{15}$N-[$^1$H, $^1$H] -TOCSY. |
| 4. | CS-ROSETTA [73–75] | 1. Structure prediction using chemical shift data from NMR, NOE distance restrains, residual dipolar coupling (RDC), and pseudo contact shifts (PCS).<br>2. The tool uses SPARTA-based selection of protein fragments from the PDB database, beyond ROSETTA Monte Carlo assembly and relaxation procedure. |

## 5.7  Molecular model visualization softwares

The different visualization software packages include MOLMOL [76], Pymol [77], Jmol/JSmol [78,79], UCSF Chimera X [80], etc. with their own set of advantages and limitations.

## 5.8  Examples

### 5.8.1  Isolated proteins

The general appearance of one-dimensional protein spectrum is shown in Fig. 5.11.

Firstly, observe two set of characteristic peaks in the negative region (<0 ppm) and dispersion of signals in the chemical shift of amidic hydrogen (~8.0−9.5 ppm). These regions are significant and reveal the native state of the globular protein. Negative signals usually refers to aliphatic groups neighboring aromatic rings, which due to the field effect of these rings present in hydrophobic pockets lead to the artificial effect of "super shielding" leading the aliphatic signals to negative values. Another parameter of critical evaluation is the dispersion rate and the width of the signals in the amide region. The signals of this region represent the slightly different chemical environments experienced by each peptide bond in a three-dimensional protein. A good signal dispersion indicates the presence of amino acid residues located in a well folded protein [45]. On the other hand, some proteins such as IDP (intrinsically unstructured proteins) shows low dispersion signals in the amide resonance region of the spectrum (caused by the similar exposure to the solvent of the peptide bonds in random structures). The absence of signals located in negative regions (<0 ppm) represents other essential spectral feature of unfolded proteins. More clearly, the chemical shift intensities around 8.3 ppm are representative of disordered proteins because this is contributed majorly by the backbone amides randomly structured. Dispersion of signals ranging from 8.5 to 11 ppm are indicative of globular folded protein structure.

The most traditional 2D NMR experiment applied for proteins analysis is HSQC (heteronuclear single quantum coherence spectroscopy), more specifically the 2D [$^1$H-$^{15}$N] HSQC (Figs. 5.12 and 5.13) and the 2D [$^1$H-$^{13}$C] HSQC. These experiments are crucial in the structural analysis of proteins to establish a single fingerprint for a given protein sample and for interaction studies with proteins or small molecules, stability studies against many variables such as temperature, chaotropic agents, salinity, pH, high pressure, or the combination of those [81].

Fig. 5.12, for instance, shows two resonance signals above 10 ppm. In principle, these signals could be indicative of indicative of belong to the Hε of the side chain of tryptophan residues. However, in this protein, there is only one tryptophan residue in the primary sequence (highlighted in red in Fig. 5.12A). This example reflects that all NMR analysis should be integrated with other spectral information. Moreover, these traditional positions described can serve as a structural probe for the researcher who wants to monitor a specific set of residues and use them as a reference for mapping interactions. HSQC experiments provide valuable information related to the overal 3D structure. Any shift or alteration on the pattern of distribution can be

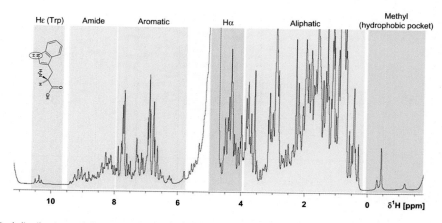

**Figure 5.11** Typical distributions of the NMR signals related to protein and its meaning in the 1D $^1$H-NMR spectrum [45].

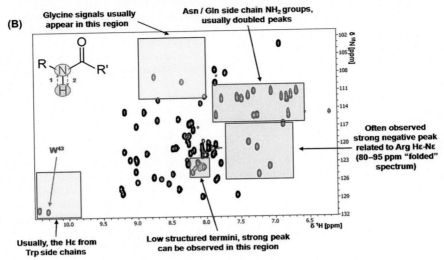

**Figure 5.12** Typical distributions of NMR peaks in [$^1$H, $^{15}$N] HSQC experiment. (A) Primary sequence of the GB1 protein, highlighted with the *red arrow*, is the tryptophan residue 43, which has the nitrogen belonged to the side chain assigned on the spectrum below. (B) The [$^1$H, $^{15}$N] HSQC spectrum of the GB1 protein [45]. These regions can be used to probe the protein and further to check the overall condition of the sample, guiding the researcher on the three-dimensional assignment.

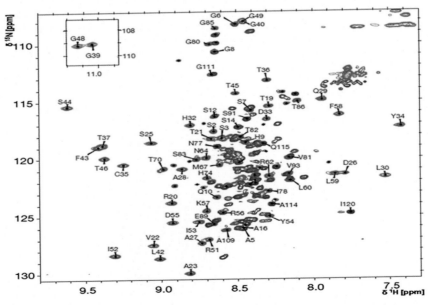

**Figure 5.13** The [$^1$H, $^{15}$N] HSQC superposition of the assigned 4E-BP2 protein. The phosphorylated protein is shown in red and nonphosphorylated is depicted in *blue*.
*Adapted from Bah A., Vernon R.M., Siddiqui Z., Krzeminski M., Muhandiram R., Zhao C., et al. Folding of an intrinsically disordered protein by phosphorylation as a regulatory switch. Nature. 2015;519(7541):106–109 [82].*

considered structural modification and be easily followed by the researcher.

It is possible to follow, for instance, Fig. 5.13 in the next session, that even an intrinsically unstructured protein (IDP) can acquire a tridimensional structure after a covalent modification. In this example, the modification caused by phosphorylation and the subsequent structural modification allows this molecule to find its substrate and play its biological role in the cell environment. The spectrum highlighted in blue shows that in this condition (WT) all the peaks belonging to the protein are almost in the same chemical environment exposed to the solvent, showing poor structured or even unstructured nature of the protein. However, considering the red spectrum (phosphorylated protein), the signal dispersion is significantly high, which shows that in this case, there are differences in the chemical environment of the H–N bond of the protein after the phosphorylation process.

## 5.8.2 Protein interactions

To understand protein interactions, the nature of these interactions should be carefully considered by the user. If there is real interaction in the proposed system, a high affinity interaction is possible that can be detected by NMR. Otherwise, the researcher has to find a better condition to stabilize the interactions by adding chemicals or altering the concentrations. Consider perform complementary approaches such as fluorescence, surface plasmon resonance (SPR), or isothermal titration calorimetry (ITC) for prior screening of interactions to define the better ratio concentration of partners in the solution. Thus, the example in Fig. 5.14 shows a high affinity interaction of lipid attachment with lipid binding protein.

The black arrows in the Fig. 5.14 indicate peak path during the increment of the ligand concentration against a fixed concentration of protein. Taking into account the high number of overlapped signals, the identification of interaction sites during the ligand titration may be difficult. In biological terms, it means that several specific residues of the protein are in direct interaction and thus there exists a possibility of one or more binding sites. An integrated analysis of the complete residue protein assignment association with the three-dimensional model, makes possible the clear identification of the set of amino acid residues that are accurately present in the interaction interface. The Fig. 5.15, shows important analysis and approaches used to determine a

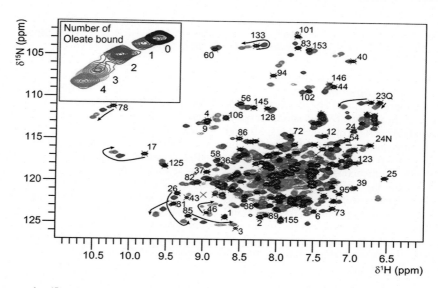

**Figure 5.14** Overlay of [$^{1}$H, $^{15}$N]-HSQC experiments related to an oleate titration against a fixed concentration of Na-FAR-1 protein. The apo (free) form of the protein is the dark blue spectrum. From purple to red color, the addition of ligand reached the ratio of 10:1. The arrows indicate the chemical shift variation during the experiment. The subset shows the precise stoichiometry of the addition and the shift of the signal of the Gly78 residue.

*Adapted from Rey-Burusco M.F., Ibanez-Shimabukuro M., Gabrielsen M., Franchini G.R., Roe A.J., Griffiths K., et al. Diversity in the structures and ligand-binding sites of nematode fatty acid and retinol-binding proteins revealed by Na-FAR-1 from Necator americanus. Biochem J. 2015;471 (3):403–414 [83].*

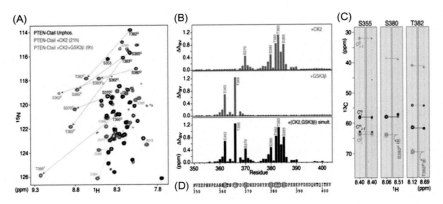

**Figure 5.15** Identification and chemical shift perturbation measurement of the PTEN-Ctail phosphosites. (A) Overlay of [¹H-¹⁵N] HSQC spectra of u-¹⁵N/¹³C PTEN-Ctail (120 μM, pH 6.7, 25°C): The nonphosphorylated PTEN-Ctail is represented in *black*. The nonphosphorylated PTEN-Ctail added to CK2 kinase during 21 h is represented in *red* and 9 h after GSK3b kinase, added to the same phosphorylated sample, is represented in *blue*. The *black* labels represent resonance assignment of the nonphosphorylated Ser/Thr residues in the spectrum. The CK2-phosphorylated Ser/Thr in *red* and GSK3b-phosphorylated is shown in *blue*. Shifted neighboring resonances are in *gray*. (B) Average (¹H, ¹⁵N) chemical shift changes, induced upon sequential phosphorylation by CK2 after 21 h (top, *red*), by GSK3b after 9 h on a CK2-phosphorylated sample (middle, *blue*) and by simultaneous phosphorylation by the two kinases (bottom, *black*). The ¹H and ¹⁵N chemical shifts of Ser/Thr residues phosphorylated by CK2 and GSK3b are labeled in *red* and *blue*, respectively. (C) ¹H, ¹³C strips from 3D-HNCACB experiments recorded on nonphosphorylated PTEN-Ctail (*gray strips*) and CK2-phosphorylated PTEN-Ctail (*pink strips*) for S355, S380, and T382. The Ca and Cb peaks are in *black* and *green*, respectively. The lines and dotted lines denote ¹³C resonances of the intra residue "i" and previous residue "i + 1," respectively. A significant downfield shift (around 2−3 ppm) is present for the Cb resonance of phosphorylated residues (e.g., S380 and T382, *red arrows*) but not for the nonphosphorylated ones (e.g., S355). (D) PTEN-Ctail sequence and CK2 (*red*) and GSK3b (*blue*) bona fide phosphosites.

*Adapted from Cordier F., Chaffotte A., Terrien E., Préhaud C., Theillet F.-X., Delepierre M., et al. Ordered phosphorylation events in two independent cascades of the PTEN C-tail revealed by NMR. J Am Chem Soc. 2012;134(50):20533−20543 [84].*

protein-nucleic acid interaction in a traditional series of [¹H, ¹⁵N]-HSQC experiments overlaid.

In Fig. 5.15A, the black [¹H, ¹⁵N]-HSQC spectrum represent the free state of the protein. The red spectrum represents the phosphorylated protein by protein kinase II after 21 hours, and the blue spectrum shows the same condition with protein kinase II and added with another protein kinase (GSK3β) simultaneously after 9 hours. The analysis shows that phosphorylation causes a significant perturbation of the chemical shifts. This protein (PTEN-Ctail) is so, substrate of the protein kinases (CK2 and GSK3 β) and exhibit a preference towards the residues of Serine and Threonine present in different positions. To quantitatively assess the process, the following formula was used for ¹H and ¹⁵N average chemical shift changes:

$$\Delta\delta_{av} = [(\Delta\delta H)^2 + (\Delta\delta N \times 0.159)^2]^{1/2}.$$

The quantitative data of perturbation after phosphorylation reflects that the residues outside the phosphorylation site have undergone structural alterations resulting in a cascade of several intracellular signaling processes, activation, or inhibition of metabolic pathways and cellular physiology, depending on the case and the biological investigation. Finally, in the analysis of the stripes of residues S355, S380, and T382 from the 3D-HNCACB spectrum, each stripe contains information about the chemical shifts of α- and β-carbons of the residue called "i" (usually more intense signals) and its predecessor "i−1" (usually less intense signals). The author, in this case, evidences the presence of the covalent modification by the alteration in the chemical shift experienced by the β carbon atom (closest to the phosphorylation site) as compared with the same structure in the absence of phosphorylation. A classical approach used for identification of residues that constitute the interaction surfaces is the chemical shift perturbation. Another tool studies the cross-relaxation of protons [85,86]. The principle of this technique is to center the saturation transference frequency in a specific atom of interest on one partner (imino protons in nucleic acids is the usual choice). The magnetic saturation diffuses through all the spins around, which leads to the decrease in the intensity of the signal of some nuclei thus assisting in the identification of spatial proximity [87,88] (Fig. 5.16).

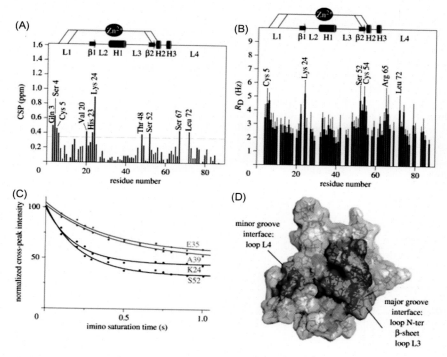

**Figure 5.16** Imino cross saturation, interaction surface mapping, and chemical shift perturbation (CSP) approach used to find the surface interaction of the THAP domain of the hTHAP1 protein and a DNA molecule. (A) Histogram of the normalized CSP observed upon DNA binding as a function of the residue number. (B) Imino cross-saturation rates (RD) as a function of residue number. (C) Examples of experimental points and fitted curves of the imino cross-saturation data. Experimental points and fitted curves are in blue for the α-helical residues (away from DNA) and in black for β-sheet residues (close to DNA). (D) Mapping of the interaction surface on the solution structure of the THAP domain of hTHAP1.

*Adapted from Campagne S., Gervais V., Milon A. Nuclear magnetic resonance analysis of protein-DNA interactions. J R Soc Interface. 2011;8 (61):1065−1078.*

The interaction between the $C_2CH$ protein THAP zinc finger domain of human THAP1 with a specific DNA target molecule is shown in Fig. 5.16. The surface of interaction presented in Fig. 5.16D was drawn as a result analysis of chemical shifts perturbation (CSP) of the protein in the presence of DNA ligand. In Fig. 5.16A, the values above the average (gray line), the N-terminal of the protein presents higher values of CSP, indicating relevant modifications on the chemical environment of these residues and possibly the site of interaction with DNA molecule. The next step shows, the cross-saturation experiment applied to a [$^1$H, $^{15}$N]-HSQC aiming to achieve the decreasing of the values of intensities caused by the cross-transfer of saturation between the molecules involved in the interaction. As mentioned earlier, the nucleic acid imino proton resonances (T—H$^3$ and G—H$^1$, $\delta^1$H centered in 13 ppm) are used to perturb the neighbor resonances in interaction in order to map the DNA—protein interactions. In Fig. 5.16B

shows the imino cross-saturated rates ($R_D = 1/T_D$; $T_D$ = decreasing time) from a mono-exponential data (Fig. 5.16C), where the normalizing decreasing intensities are exposed versus saturation time. An expressive match between these two techniques, allow the researcher have a reliable identification. In order to visualize this data, a 3D model is highlighted in the Fig. 5.16D [89]. The continuous surface pattern observed is desirable for two macromolecule interaction. The combination of these approaches are powerful tool to screen and identify the surface of interaction of protein—protein or protein—DNA interaction.

### 5.8.3    Nucleic acids (DNA and RNA)

With the advent of DNA chemical synthesis, the development of more accurate methods of spectral acquisition and isotopic labeling was possible [90]. An important experiment often used by structural biologists is mapping the interactions of nucleic acid with proteins. An intrinsic feature of the nucleic acid molecules is the intermediate flexibility, once that this molecule has to be capable to diffuse within the cytoplasm and at the same time conserve some structure to be capable to interact with its molecular partner [6]. Significant advances have been made in the development of pulse sequences and experiments to overcome

inherent problems related to nucleic acids in molecular complexes such as the reduced number of intermolecular NOEs signals and size of the complexes [91,92].

## 5.8.4 General overview

NMR for the study of nucleic acid has advanced our understanding related to the secondary structure of DNA and RNA, their biochemical roles, and drug designing [93]. Figs. 5.17 and 5.18 show a 1D spectrum where the

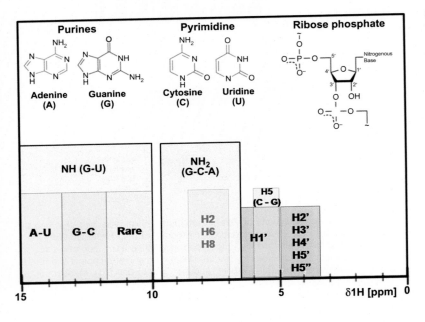

**Figure 5.17** Typical distributions of the NMR signals related to RNA molecule analysis and its identifications at $^1$H unidimensional experiment [45].

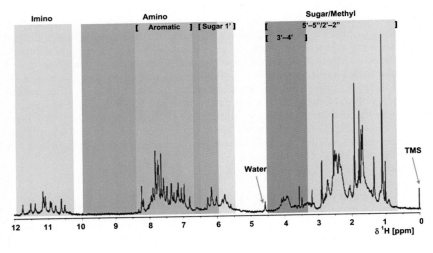

**Figure 5.18** Typical distributions of the NMR peaks in DNA molecule and its identifications at 1D $^1$H-NMR spectrum [45].

usual location of the chemical groups present in nucleic acids are depicted [45]. In the double stranded DNA molecule, it is possible to monitor interactions between nitrogenous bases. The signals belonging to the imino groups ($\sim 10-12$ ppm) change noticeably and are useful for probing three-dimensional perturbations [94,95]. The intermolecular interactions between nitrogenous bases and protons belonging to ribose show greater modifications when in complex with macromolecules like proteins or other nucleic acids, which is useful to monitor interactions. Thus, the region comprising signals generally below 7 ppm may be useful for overall assessment of interaction sites [45,96].

## 5.9 Conclusion

The NMR technique has brought an unparalleled contribution to the biomolecular studies, both related to structural elucidation and the advancement in dynamics studies. A detailed study of enzymatic catalysis, carbohydrate dynamics, and protein folding dynamics is made feasible. The interaction between molecules and their biological targets, as well as understanding about evolution of the misfolding diseases, are some of the other domains that are deciphered using this technique. We have overviewed many techniques in this chapter that can help a beginning user in the analysis of the NMR spectrum to plan their experiments for different objectives. Nevertheless, it is necessary to dedicate the handling of the magnets with specialized personnel to further data processing and analysis skills. The participation in class courses offered throughout the year by the worldwide NMR community is a very positive approach that will help the beginner NMR spectroscopist. Always have in hand theoretical materials and reference tables for appropriate data handling. These documents are of extreme importance since they are efficient and precise, concerning the challenges presented by the biopolymers analysis.

## References

[1] Henzler-Wildman K, Kern D. Dynamic personalities of proteins. Nature. 2007;450(7172):964−72.

[2] Mittermaier A. New tools provide new insights in NMR studies of protein dynamics. Science 2006; 312:224−8.

[3] Henzler-Wildman KA, Lei M, Thai V, Kerns SJ, Karplus M, Kern D. A hierarchy of timescales in protein dynamics is linked to enzyme catalysis. Nature 2007;450:913−16.

[4] Laurence TA, Kong X, Jager M, Weiss S. Probing structural heterogeneities and fluctuations of nucleic acids and denatured proteins. Proc Natl Acad Sci 2005;102: 17348−53.

[5] Miller TR, Alley SC, Reese AW, Solomon MS, McCallister WV, Mailer C, et al. A Probe for Sequence-Dependent Nucleic AcidDynamics [Internet]. 2002 [cited 2018 Aug 14]. Available from: https://pubs.acs.org/doi/abs/ 10.1021/ja00141a040

[6] Al-Hashimi HM. NMR studies of nucleic acid dynamics. J Magn Reson 2013;237:191−204.

[7] Ishima R, Torchia DA. Protein dynamics from NMR. Nat Struct Biol 2000;7(9):740−3.

[8] Tolman JR, Al-Hashimi HM, Kay LE, Prestegard JH. Structural and dynamic analysis of residual dipolar coupling data for proteins. J Am Chem Soc 2001;123(7): 1416−24.

[9] Engen JR. Analysis of protein conformation and dynamics by hydrogen/deuterium exchange MS. Anal Chem 2009;81 (19):7870−5.

[10] Jarymowycz VA, Stone MJ. Fast time scale dynamics of protein backbones: NMR relaxation methods, applications, and functional consequences. Chem Rev. 2006; 106(5):1624−71.

[11] The Nobel Prize in Physics, 1943 [Internet]. [cited 2018 Aug 14]. Available from: https://www. nobelprize.org/nobel_prizes/physics/laureates/1943/

[12] Isidor Isaac Rabi - Biographical [Internet]. [cited 2018 Aug 14]. Available from: https://www.nobel prize.org/nobel_prizes/physics/laureates/1944/rabi-bio.html

[13] The Nobel Prize in Physics, 1952 [Internet]. [cited 2018 Aug 14]. Available from: https://www. nobelprize.org/nobel_prizes/physics/laureates/1952/

[14] Richard R. Ernst - Facts [Internet]. [cited 2018 Aug 14]. Available from: https://www.nobelprize.org/ nobel_prizes/chemistry/laureates/ 1991/ernst-facts.html

[15] Kurt Wüthrich - Facts [Internet]. [cited 2018 Aug 14]. Available from: https://www.nobelprize.org/ nobel_prizes/chemistry/laureates/ 2002/wuthrich-facts.html

[16] Hong M, Zhang Y, Hu F. Membrane protein structure and dynamics from NMR spectroscopy. Annu Rev Phys Chem 2012;63 (1):1−24.

[17] Hubbell WL, López CJ, Altenbach C, Yang Z. Technological advances in site-directed spin labeling of proteins. Curr Opin Struct Biol 2013;23(5):725−33.

[18] Otting G. Protein NMR using paramagnetic ions. Annu Rev Biophys 2010;39(1):387−405.

[19] Viegas A, Manso J, Nobrega FL, Cabrita EJ. Saturation-transfer

difference (STD) NMR: a simple and fast method for ligand screening and characterization of protein binding. J Chem Educ. 2011;88 (7):990–4.

[20] Kerfah R, Plevin MJ, Sounier R, Gans P, Boisbouvier J. Methyl-specific isotopic labeling: a molecular tool box for solution NMR studies of large proteins. Curr Opin Struct Biol. 2015;32:113–22.

[21] Gans P, Hamelin O, Sounier R, Ayala I, Durá MA, Amero CD, et al. Stereospecific isotopic labeling of methyl groups for NMR spectroscopic studies of high-molecular-weight proteins. Angew Chem. 2010;122 (11):2002–6.

[22] Miyanoiri Y, Takeda M, Okuma K, Ono AM, Terauchi T, Kainosho M. Differential isotope-labeling for Leu and Val residues in a protein by E. coli cellular expression using stereo-specifically methyl labeled amino acids. J Biomol NMR 2013;57(3):237–49.

[23] Cala O, Guillière F, Krimm I. NMR-based analysis of protein–ligand interactions. Anal Bioanal Chem. 2014;406(4):943–56.

[24] Vazquez L, e Lima LMT da R, Almeida M da S. Comprehensive structural analysis of designed incomplete polypeptide chains of the replicase nonstructural protein 1 from the severe acute respiratory syndrome coronavirus. Rezaei H, editor. PLoS One 2017;12(7) e0182132.

[25] Smolinska A, Blanchet L, Buydens LMC, Wijmenga SS. NMR and pattern recognition methods in metabolomics: from data acquisition to biomarker discovery: a review. Anal Chim Acta. 2012;750: 82–97.

[26] Fidalgo TKS, Freitas-Fernandes LB, Angeli R, Muniz AMS, Gonsalves E, Santos R, et al. Salivary metabolite signatures of children with and without dental caries lesions. Metabolomics. 2013;9(3):657–66.

[27] El-Bacha T, Struchiner CJ, Cordeiro MT, Almeida FCL, Marques ET, Da Poian AT. 1H nuclear magnetic resonance metabolomics of plasma unveils liver dysfunction in dengue patients. Ou J-HJ, editor. J Virol 2016;90 (16):7429–43.

[28] Breton RC, Reynolds WF. Using NMR to identify and characterize natural products. Nat Prod Rep. 2013;30:501.

[29] Kitayama T, Hatada K. NMR spectroscopy of polymers. Springer Science & Business Media; 2013. 231 p.

[30] Slonim IY. The NMR of polymers. Springer Science & Business Media; 2012. 366 p.

[31] Gardner KH, Kay LE. The use of 2 H, $1^3$ C, $1^5$ N multidimensional NMR to study the structure and dynamics of proteins. Annu Rev Biophys Biomol Struct. 1998;27 (1):357–406.

[32] Aue WP, Bartholdi E, Ernst RR. Two-dimensional spectroscopy. Application to nuclear magnetic resonance. J Chem Phys 1976;64 (5):2229–46.

[33] Kimmich R. NMR: tomography, diffusometry, relaxometry. Springer Science & Business Media; 2012. 545 p.

[34] Mariette F. Investigations of food colloids by NMR and MRI. Curr Opin Colloid Interface Sci. 2009; 14(3):203–11.

[35] van Duynhoven J, Voda A, Witek M, Van As H. Time-Domain NMR Applied to Food Products. In: Annual Reports on NMR Spectroscopy.

[36] The Nobel Prize in Physiology or Medicine, 2003 [Internet]. [cited 2018 Aug 14]. Available from: https://www.nobelprize.org/nobel_prizes/medicine/laureates/2003/.

[37] Hallenga K, Koenig SH. Protein rotational relaxation as studied by solvent proton and deuteron magnetic relaxation. Biochemistry. 1976;15(19):4255–64.

[38] Rabi II. Space quantization in a gyrating magnetic field. Phys Rev. 1937;51(8):652–4.

[39] Carr HY, Purcell EM. Effects of diffusion on free precession in nuclear magnetic resonance experiments. Phys Rev. 1954;94(3):630–8.

[40] Jacobsen NE. NMR spectroscopy explained: simplified theory, applications and examples for organic chemistry and structural biology. John Wiley & Sons; 2007. 687 p.

[41] Markley JL, Bax A, Arata Y, Hilbers CW, Kaptein R, Sykes BD, et al. Recommendations for the presentation of NMR structures of proteins and nucleic acids – IUPAC-IUBMB-IUPAB Inter-Union Task Group on the Standardization of Data Bases of Protein and Nucleic Acid Structures Determined by NMR Spectroscopy. J Biomol NMR. 1998;12(1):1–23.

[42] Venters R, Huang C-C, Farmer B, Trolard R, Spicer L, Fierke C. High-level 2H/13C/15N labeling of proteins for NMR studies. J Biomol NMR [Internet]. 1995;5(4).

[43] Marley J, Lu M, Bracken C. A method for efficient isotopic labeling of recombinant proteins. J Biomol NMR. 2001;20(1): 71–5.

[44] Keeler J. Understanding NMR spectroscopy. John Wiley & Sons; 2011. 676 p.

[45] Wüthrich K. NMR of proteins and nucleic acids. Wiley; 1986. 336 p.

[46] Levitt MH. Spin dynamics: basics of nuclear magnetic resonance. John Wiley & Sons; 2001. 718 p.

[47] Kwan AH, Mobli M, Gooley PR, King GF, Mackay JP. Macromolecular NMR spectroscopy for the non-spectroscopist: macromolecular NMR for the non-spectroscopists I. FEBS J. 2011;278 (5):687–703.

[48] Andrew ER. Magic angle spinning in solid state n.m.r. spectroscopy. Philos Trans R Soc Math Phys Eng Sci. 1981;299(1452):505–20.

[49] Opella SJ. Solid state NMR of biological systems. Annu Rev Phys Chem. 1982;33(1):533–62.

[50] Opella SJ, Hexem JG, Frey MH, Cross TA, Derbyshire W. Solid state n.m.r. of biopolymers. Philos Trans R Soc Math Phys Eng Sci. 1981;299(1452):665–83.

[51] Structure Determination of Organic Compounds [Internet]. Berlin, Heidelberg: Springer Berlin Heidelberg; 2009.

[52] Silverstein RM, Bassler GC. Spectrometric identification of organic compounds. J Chem Educ. 1962;39(11):546.

[53] Bauer CJ, Frenkiel TA, Lane AN. A comparison of the ROESY and NOESY experiments for large molecules, with application to nucleic acids. J Magn Reson 1969 1990;87(1):144–52.

[54] Mumenthaler C, Güntert P, Braun W, Wüthrich K. Automated combined assignment of NOESY spectra and three-dimensional protein structure determination. J Biomol NMR 1997;10(4):351–62.

[55] Mo H, Pochapsky TC. Intermolecular interactions characterized by nuclear Overhauser effects. Prog Nucl Magn Reson Spectrosc. 1997;30(1–2):1–38.

[56] Acton TB, Xiao R, Anderson S, Aramini J, Buchwald WA, Ciccosanti C, et al. Preparation of protein samples for NMR structure, function, and small-molecule screening studies. Methods in enzymology. Elsevier; 2011. p. 21–60 [cited2018 Aug 14].

[57] Englander SW, Mayne L. Protein folding studied using hydrogen-exchange labeling and two-dimensional NMR. Annu Rev Biophys Biomol Struct. 1992;21 (1):243–65.

[58] Hvidt A, Nielsen SO. Hydrogen exchange in proteins. Advances in protein chemistry [Internet]. Elsevier; 1966. p. 287–386.

[59] TopSpin 3.5. ® 2017. Bruker BioSpin GmbH.

[60] Marat K. SpinWorks software. University of Manitoba; 2005.

[61] Vranken WF, Boucher W, Stevens TJ, Fogh RH, Pajon A, Llinas M, et al. The CCPN data model for NMR spectroscopy: development of a software pipeline. Proteins 2005;59:687–96.

[62] Keller RLJ. Optimizing the process of nuclear magnetic resonance spectrum analysis and computer aided resonance assignment [Internet]. ETH Zurich; 2005.

[63] Lee W, Tonelli M, Markley JL. NMRFAM-SPARKY: enhanced software for biomolecular NMR spectroscopy. Bioinformatics. 2015;31 (8):1325–7.

[64] Vranken WF, Boucher W, Stevens TJ, Fogh RH, Pajon A, Llinas M, et al. The CCPN data model for NMR spectroscopy: development of a software pipeline. Proteins Struct Funct Bioinforma. 2005;59 (4):687–96.

[65] Delaglio F, Grzesiek S, Vuister G, Zhu G, Pfeifer J, Bax A. NMRPipe: a multidimensional spectral processing system based on UNIX pipes. J Biomol NMR 1995;6(3).

[66] Hyberts SG, Arthanari H, Wagner G. Applications of non-uniform sampling and processing. In: Billeter M, Orekhov V, editors. Novel sampling approaches in higher dimensional NMR. Berlin, Heidelberg: Springer Berlin Heidelberg; 2011. p. 125–48.

[67] Linge JP, Habeck M, Rieping W, Nilges M. ARIA: automated NOE assignment and NMR structure calculation. Bioinformatics. 2003;19: 315–16.

[68] Güntert P. Automated NMR structure calculation with CYANA. Protein NMR techniques. New Jersey: Humana Press; 2004. p. 353–78.

[69] Serrano P, Pedrini B, Mohanty B, Geralt M, Herrmann T, Wüthrich K. The J-UNIO protocol for automated protein structure determination by NMR in solution. J Biomol NMR. 2012;53(4):341–54.

[70] Volk J, Herrmann T, Wüthrich K. Automated sequence-specific protein NMR assignment using the memetic algorithm MATCH. J Biomol NMR. 2008;41(3):127–38.

[71] Fiorito F, Herrmann T, Damberger FF, Wüthrich K. Automated amino acid side-chain NMR assignment of proteins using 13C- and 15N-resolved 3D [1H,1H]-NOESY. J Biomol NMR. 2008;42(1):23–33.

[72] Herrmann T, Güntert P, Wüthrich K. Protein NMR structure determination with automated NOE assignment using the new software CANDID and the torsion angle dynamics algorithm DYANA. J Mol Biol. 2002;319(1):209–27.

[73] Herrmann T, Güntert P, Wüthrich K. Protein NMR structure determination with automated NOE-identification in the NOESY spectra using the new software ATNOS. J Biomol NMR. 2002;24(3):171–89.

[74] Raman S, Lange OF, Rossi P, Tyka M, Wang X, Aramini J, et al. NMR structure determination for larger proteins using backbone-only data. Science. 2010;327 (5968):1014–18.

[75] Shen Y, Lange O, Delaglio F, Rossi P, Aramini JM, Liu G, et al. Consistent blind protein structure generation from NMR chemical shift data. Proc Natl Acad Sci. 2008;105(12):4685–90.

[76] Koradi R, Billeter M, Wüthrich K. MOLMOL: a program for display and analysis of macromolecular structures. J Mol Graph. 1996;14 (1):51–5.

[77] The PyMOL Molecular Graphics System. LLC.

[78] Shenelle Pearl Ghulam JP. Introduction to molecular modeling in chemistry education. e-Oppi Ltd. & Edumendo Publishing; 2017. 120 p.

[79] Gutow JH. Easy jmol web pages using the jmol export to web function: a tool for creating interactive web-based instructional resources and student projects with live 3-D

images of molecules without writing computer code. J Chem Educ. 2010;87(6):652–3.

[80] Pettersen EF, Goddard TD, Huang CC, Couch GS, Greenblatt DM, Meng EC, et al. UCSF Chimera? A visualization system for exploratory research and analysis. J Comput Chem. 2004;25(13):1605–12.

[81] Foguel D, Silva JL. New insights into the mechanisms of protein misfolding and aggregation in amyloidogenic diseases derived from pressure studies. Biochemistry. 2004;43(36):11361–70.

[82] Bah A, Vernon RM, Siddiqui Z, Krzeminski M, Muhandiram R, Zhao C, et al. Folding of an intrinsically disordered protein by phosphorylation as a regulatory switch. Nature. 2015;519(7541):106–9.

[83] Rey-Burusco MF, Ibanez-Shimabukuro M, Gabrielsen M, Franchini GR, Roe AJ, Griffiths K, et al. Diversity in the structures and ligand-binding sites of nematode fatty acid and retinol-binding proteins revealed by Na-FAR-1 from *Necator americanus*. Biochem J. 2015;471(3):403–14.

[84] Cordier F, Chaffotte A, Terrien E, Préhaud C, Theillet F-X, Delepierre M, et al. Ordered phosphorylation events in two independent cascades of the PTEN C-tail revealed by NMR. J Am Chem Soc. 2012;134(50):20533–43.

[85] Williamson MP. Using chemical shift perturbation to characterise ligand binding. Prog Nucl Magn Reson Spectrosc. 2013;73:1–16.

[86] Williamson MP. Chemical shift perturbation. In: Webb GA, editor. Modern magnetic resonance. Cham: Springer International Publishing; 2017. p. 1–19.

[87] Akasaka K. Longitudinal relaxation of protons under cross saturation and spin diffusion. J Magn Reson 1969 1981;45(2):337–43.

[88] Shimada I. NMR techniques for identifying the interface of a larger protein–protein complex: cross-saturation and transferred cross-saturation experiments. Methods in enzymology. Elsevier; 2005. p. 483–506.

[89] Campagne S, Gervais V, Milon A. Nuclear magnetic resonance analysis of protein-DNA interactions. J R Soc Interface. 2011;8(61):1065–78.

[90] Lu K, Miyazaki Y, Summers MF. Isotope labeling strategies for NMR studies of RNA. J Biomol NMR. 2010;46(1):113–25.

[91] Hennig M, Williamson JR, Brodsky AS, Battiste JL. Recent advances in RNA structure determination by NMR. In: Beaucage SL, Bergstrom DE, Herdewijn P, Matsuda A, editors. Current protocols in nucleic acid chemistry [Internet]. Hoboken, NJ: John Wiley & Sons, Inc; 2000. p. 7.7.1–7.7.30.

[92] Göbl C, Madl T, Simon B, Sattler M. NMR approaches for structural analysis of multidomain proteins and complexes in solution. Prog Nucl Magn Reson Spectrosc. 2014;80:26–63.

[93] Zerbe O. BioNMR in drug research. John Wiley & Sons; 2006. p. 515.

[94] Wemmer DE, Chou SH, Hare DR, Reid BR. Duplex-hairpin transitions in DNA: NMR studies on CGCGTATACGCG. Nucl Acids Res. 1985;13(10):3755–72.

[95] Chou SH, Wemmer DE, Hare DR, Reid BR. Sequence-specific recognition of DNA: NMR studies of the imino protons of a synthetic RNA polymerase promoter. Biochemistry. 1984;23(10):2257–62.

[96] Russu I. Studying DNA-protein interactions using NMR. Trends Biotechnol. 1991;9(1):96–104.

# Chapter 6

# Fourier transform infrared spectroscopy: Data interpretation and applications in structure elucidation and analysis of small molecules and nanostructures

Atul Kumar[1], Mahima Khandelwal[2], Sudhir K. Gupta[3], Vinit Kumar[4] and Reshma Rani[5]

[1]Amity Institute of Engineering & Technology, Amity University, Greater Noida, India, [2]School of Chemical Engineering, University of Ulsan, South Korea, [3]Faculty of Chemical Sciences, Department of Chemistry, Harcourt Butler Technical University (Formerly HBTI), Kanpur, India, [4]Amity Institute of Molecular Medicine and Stem Cell Research, Amity University, Noida, India, [5]Amity Institute of Biotechnology, Amity University, Noida, India

## 6.1 Introduction

Infrared spectroscopy (IR spectroscopy) is an important technique that deals with the interaction of a molecule with IR range of the electromagnetic spectrum ranging from 4000 to 400 cm$^{-1}$. The necessary condition for a molecule or sample to show infrared spectrum is the change in the electric dipole moment of the functional group present in a molecule or a sample during the vibration based on the selection rule for IR transitions [1,2]. Molecules whose dipole moments change during vibration are infrared-active. IR spectroscopy primarily helps in the identification of the type of chemical bonds present in a sample that are reflected by the absorption of characteristic wavelength of the infrared radiation responsible for the vibrational transition induced from particular functional groups [3]. The great advantage of infrared spectroscopy is that samples in the form of liquid, gas, solid, powder, or film can all be studied with a careful selection of sampling technique. The more advanced technique Fourier transform infrared (FTIR) spectroscopy offers a facile kinetic and mechanistic insight of the chemical functionalities responsible for any chemical change or a noncovalent supramolecular interaction. These vibrational frequencies can be analyzed easily and

quickly through high-resolution spectra for both solid and liquid samples, and that can be done nondestructively. The variations in the shape of characteristic peaks/bands along with their intensity are crucial in predicting the chemical environment of the functionalities present in the vicinity and dynamic changes occurring therein, which unveils the kind of functionalities or ligand with which the samples are interacting [4–6]. IR spectroscopy has been extensively used for qualitative as well as quantitative analysis in academic labs as well as in industry. The extent of application of IR spectroscopy for achieving a deeper insight of structural analysis and interactions at molecular level has been intensive with the instrumental advancements. Concisely, IR spectroscopy is a promising tool in (1) establishing the chemical structure of small molecules, natural products, and other biomolecules; (2) identification of functional groups in sample; and (3) identification and characterization of supramolecular interactions and chemical bonding in supramolecular chemistry [7–10]. Therefore, IR spectroscopy not only offers various applications in organic chemistry, drug discovery, and drug design but also provides valuable information in a comprehensive mechanism of the morphological transitions within the phase(s) of metal/metal oxides, metal nanoparticles, carbon nanoparticles and graphene quantum dots (GQDs)/sheet along with their interactions with biomolecules (Fig. 6.1) [11,12]. A few significant reports have been discussed in this chapter wherein IR spectroscopy played a significant role to investigate the chemical structure, functional group present in chemical entity, nature of interactions, structural transformations, and interactions of different biomolecules with nanoparticles.

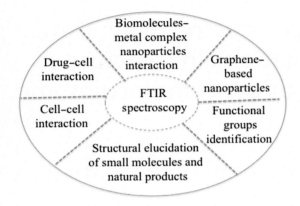

**Figure 6.1** Wide applications of FTIR in different research areas.

## 6.2 Interpretation of biorelevant small molecules

IR spectroscopy is an important tool to elucidate the chemical structure of newly synthesized small molecules. In fact, this structure elucidation depends mainly on three facts: (1) wavenumber, which reflects the position of absorbance and depends on the energy required for absorbance; (2) intensity of the absorbance peak, which is related to dipole/strength of bond present in the molecules; (3) shape of IR band may be broad or sharp, which provides information about the type of bonds. There is a large class of biorelevant molecules discovered so far whose structure has been examined with the help of IR spectra by correlating the various bands that arise in spectrum of a particular compound with functional group present. For example, small molecules such as cyclic-imides (1−4) [13], spiro-based imides (5, 6) [14,15], amidine (7,8) [16], and bis-amidines (9) [17] (Fig. 6.2) were examined by IR and the characteristic IR bands at different wavenumber were found in correlation with the functional group present in the molecules. Overall, all the imides compounds including cyclic-imides (1−4), spiro-imide (5), and bis-imide (6) displayed a common

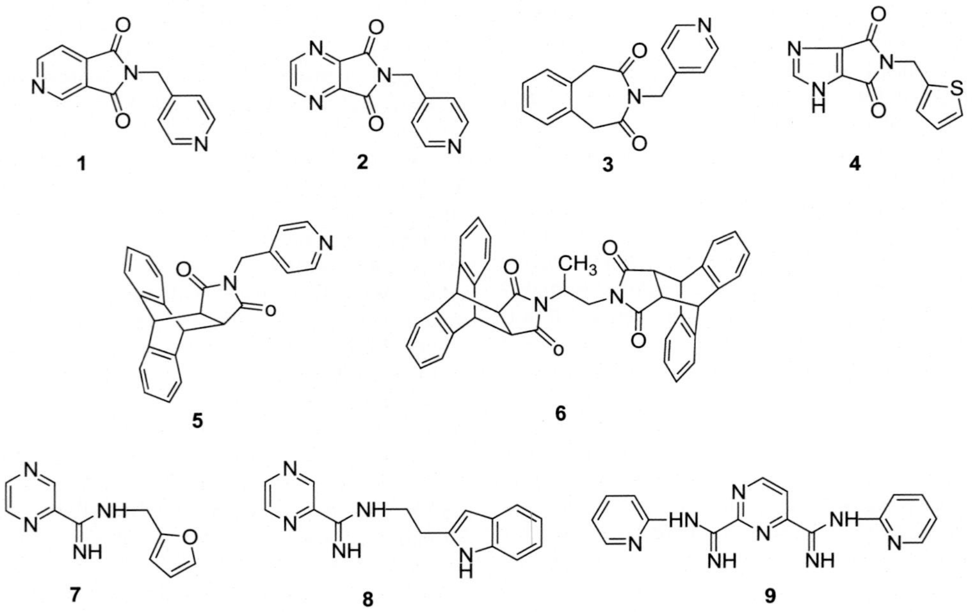

1

2

3

4

5

6

7

8

9

**Figure 6.2** Chemical structure of cyclic-imides (1−4), spiro-imides (5), bis-imides (6), amidines (7,8), bis-amidine (9).

structural feature, which is cyclic —CO—N—CO—bond characterized by two IR peaks observed in the range of 1700−1740 and 1750−1800 cm$^{-1}$.

The vibration bands in the range of 1515−1588 and 1447−1489 cm$^{-1}$ in all the compounds mentioned in Fig. 6.2 were assigned to aromatic group [13−15]. Bis-imide (6) (Fig. 6.2) exhibited strong IR absorption band at 1780 and 1660 cm$^{-1}$ assigned to —CO—N—CO—functional group, 1571 and 1463 cm$^{-1}$ assigned for aromatic region. The absence of band at ≈ 3333 cm$^{-1}$ confirms the absence of NH$_2$ and NH group that was present in the reactant. Further, the amidine (7,8) and bis-amidine (9) (Fig. 6.2) both showed characteristic peaks at 1447−1489 cm$^{-1}$ assigned to NH and —C═NH functional group [16,17]. In some cases, these amidine type chemical structures also showed absorption peak at ≈ 3433 cm$^{-1}$ assigned to —NH$_2$ functional group due to resonating structure. All these molecules were found to exhibit very good antiinflammatory activities as reported in Table 6.1. IR data of all the cyclicimides (1−4), spiro-imide (5), and bis-imide (6) are listed in Table 6.1.

Acridine derivatives have been synthesized with broad spectrum of biological activities. Due to the planar structure, acridine derivatives (11,12,13) endowed strong anticancer activities via intercalation with DNA base pairs ultimately resulting in cell cycle arrest and apoptosis [18,19]. In literature, the structural elucidation of majority of acridine derivatives has been done by IR spectroscopy. For example, the characteristic IR absorption band at 3459 and 3416 cm$^{-1}$ assigned to —NH—functional group present in compound 10 and 11 (Fig. 6.3) whereas in

**Table 6.1** IR data of all the cyclic-imides (1−4), spiro-imide (5), and bis-imide (6).

| Sr. no. | Compound no. | Antiinflammatory activity (50 mg kg$^{-1}$ po) | IR data (KBr) (in cm$^{-1}$) |
|---|---|---|---|
| 1 | 1 | 33 | $\nu_{max}$: 1697 (—CO—N—CO—), 1564, and 1489 (Ar) |
| 2 | 2 | 32 | $\nu_{max}$: 1638 (—CO—N—CO—), 1588, and 1447 (Ar) |
| 3 | 3 | 40 | $\nu_{max}$: 1728 (—CO—N—CO—), 1564, and 1421 (Ar) |
| 4 | 4 | 34 | $\nu_{max}$: 1703 (—CO—N—CO—), 1646 (—C═N—), 1544, and 1483 (Ar) |
| 5 | 5 | 35 | $\nu_{max}$: 1698 (—CO—N—CO—), 1462, and 1389 (Ar) |
| 6 | 6 | 22 | $\nu_{max}$: 1780, 1660 (—CO—N—CO—), 1571, and 1463 (Ar) |
| 7 | 7 | 36 | $\nu_{max}$: 3481 (—NH—), 1643 (—C═N—), 1582, 1420, 1582, and 1420 (Ar) |
| 8 | 8 | 24 | $\nu_{max}$: 3441 (—NH—), 1651 (—C═N—), 1589, and 1435 (Ar) |
| 9 | 9 | 26 | $\nu_{max}$: 3423 (—NH—), 1634 (—C═N—), 1609, and 1455 (Ar) |

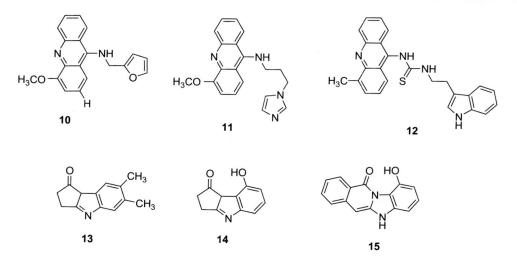

**Figure 6.3** Acridine derivatives (10−12), planar tricyclic (13, 14), and tetracyclic (15) molecules.

**Table 6.2 IR data of acridine derivatives (10−12), planar tricyclic (13, 14), and tetracyclic (15).**

| Sr. no. | Compound no. | Antiinflammatory activity (50 mg kg$^{-1}$ po) | IR data (KBr) (in cm$^{-1}$) |
|---|---|---|---|
| 1 | 10 | 38 | $\nu_{max}$: 3459 (−NH−), 1630, and 1570 (Ar) |
| 2 | 11 | 17 | $\nu_{max}$: 3416 (−NH−), 1630, and 1575 (Ar) |
| 3 | 12 | 37 | $\nu_{max}$: 3430 and 3236 (−NH−and −NH−), 1632, 1586, and 1562 (Ar) |
| 4 | 13 | 39 | $\nu_{max}$: 3129 (−NH−) 1703 (−CO−), 1630 (−C=C−), 1579, 1472, and 1415 (Ar) |
| 5 | 14 | 39 | $\nu_{max}$: 3410, 3068 (−NH−and −OH), 1705 (−CO−), 1631 (−C=C−), 1597, 1486, and 1408 (Ar) |
| 6 | 15 | 29 | $\nu_{max}$: 3138, (−NH−), 1677 (−CO−), 1576, 1491, and 1472 (Ar) |

compound **12** where −NH−CS−NH−group is present, the absorption peaks at 3430 and 3236 cm$^{-1}$ are allotted to the two NH groups of thiourea moiety present in the compound. In these compounds, other peaks at 1630, 1570, 1632, 1586, and 1562 cm$^{-1}$ were allotted to aromatic region [18,19]. Most common IR absorption peaks observed in spectra of these compounds are listed in Table 6.2.

Further, the tricyclic and tetracyclic small molecules **13**, **14**, and **15** also possessed very good absorption and fluorescence properties, further exhibited antiinflammatory as well as anticancer properties [20,21]. These compounds exhibited characteristic IR bands due to −NH, >CO, −C=N and presence of aromatic region. The IR spectrum of tricyclic compound **13** displayed strong absorption peak at 3129 cm$^{-1}$ assigned to −NH whereas absorption peak at 1703 cm$^{-1}$ (due existence of resonating structure) and 1630 cm$^{-1}$ correlated to carbonyl (−CO−) functional group and alkene functional group (−C=C−) respectively. The IR band observed at 1579, 1472, and 1415 cm$^{-1}$ was due to aromatic region. Further, compound **14** also yielded similar type of IR spectrum in which absorption band at 3068 cm$^{-1}$ was assigned to NH (due to resonating structure) and IR band at frequencies 1705 and 1631 cm$^{-1}$ were correlated to carbonyl (−CO−) and alkene (−C=C−) group respectively (Table 6.2). The IR absorption peak at 3410 cm$^{-1}$ observed due to the presence of hydroxyl group in IR spectrum of **14**. Other IR bands at 1597, 1486, 1408 cm$^{-1}$ were assigned to aromatic regions [20,21]. Similar pattern of IR spectrum of compound **15** was observed in IR spectrum of **15**, which showed IR band at 3138 cm$^{-1}$ assigned to −NH and at 1677 cm$^{-1}$ assigned to −CO−. Overall IR data of compound **13−15** listed in Table 6.2.

## 6.3 Ligand-binding interactions

In the drug development process, investigation of ligand-binding interactions with biomolecules is an essential trigger point by which time and cost can be reduced to discover a new drug. The IR spectroscopy plays significant role to examine the ligand-binding interactions at molecular level by detecting ligand induced conformational changes [22]. These structural changes depending upon binding strength yielded distinct infrared absorption signals, which ultimately reflected binding mode of ligand into the binding cavity of biomolecules [23,24]. Kumar and Barth investigated the structural transformations in pyruvate kinase (PK), an important enzyme in the glycolytic pathway, stimulated by the interaction (binding) of phosphoenolpyruvate (PEP) and magnesium ion (Mg$^{2+}$) by applying the IR spectroscopy. The PK catalyzes the physiological reaction, which is the transfer of a phosphate group from PEP to adenosine diphosphate (ADP). The positive IR peaks appear due to the attachment of substrate PEP to Mg$^{2+}$, which induces the changes in structure of the PK enzyme, however small conformational alterations were also observed in the backbone of PK enzyme. The antisymmetric stretching mode of COO$^-$ groups gave positive bands at frequency 1590 and 1551 cm$^{-1}$ while the symmetric stretching mode showed IR band at 1415 cm$^{-1}$. Along with these bands, the IR bands at 1214 cm$^{-1}$ allotted to stretching vibration mode of C−O and IR peaks at 1124 and 1110 cm$^{-1}$ correlate asymmetric stretching vibration of the PO$_3^{-2}$ functional group, whereas symmetric stretching vibration of PO$_3^{-2}$ gave rise to the band at 967 cm$^{-1}$. These IR bands provide information about the geometry changes induced by binding of PEP and change in bond strengths of carboxylate and phosphate groups [25].

## 6.4 Drug—cell interactions

IR spectroscopy has attracted much attention in monitoring the drug—cell interactions and cellular response towards drugs in vitro particularly in cancer cells and in examination of valuable changes in the spectral signatures of cells (cellular response) in the retort of anticancer drugs. Responses of several anticancer agents on cancer cells have been studied by IR spectroscopy. For example, effect of anticancer agent namely cardiotonic steroids was investigated on prostate cancer cell lines by IR spectroscopy [26]. Later on, effect of naturally occurring polyphenols curcumin, gallate (EGCG), and epigallocatechin on exposure to cancer cells in vitro was also analyzed successfully by FTIR spectroscopy [27]. Platinum compounds have been explored to examine their effect on cyclooxygenase-2 (COX-2) expression in breast cancer and a close association of CacyBp downregulation with poor prognosis in breast cancer was observed. [28,29] Moreover, the scope of IR spectroscopy is not only limited to (1) structure elucidation of small molecules and drug-like molecules, (2) drug designing and drug discovery, (3) identification of natural products, (4) study ligand-binding interaction, and (5) examining drug—cell interactions but is also useful in the study of biomolecules—metal interactions, metal—metal interactions and biomolecules—inorganic nanoparticles hybrids.

## 6.5 Biomolecules metal/metal complexes

The interactions of biomolecules with metal/metal complexes have been fundamental in sustaining the energy cycle and life on the Earth. In living beings, metals and biomolecules both are needed in various forms. Existence of life under physiological conditions has been feasible due to the synergistic interactions of metals forming complexes with bioligands [30]. The metals are not only involved in the formation of specific biomolecular structures but also significantly influence the organization of biomolecules through supramolecular interactions. Due to the tremendous advancements in technology and instrumentation over the years, the chemistry of many bioinorganic systems has been investigated. However, there are mechanisms and interactions that are still not completely understood such as complete mechanism of photosynthesis, the role of V and Se in life. There are number of reports available on biomineralization wherein the synthesis of inorganic materials with superior properties based on the interaction of proteins and peptides with

the inorganic materials within organisms have been addressed [31]. Among all the metals, iron is one of the most profuse metals in the earth's crust and has been an integral part of life on Earth since the very beginning, as iron plays key role in (1) primitive *Clostridium* bacteria for electron transfer reaction; (2) oxygen transfer protein, that is, hemoglobin; (3) oxygen storage protein, that is, myoglobin, ferritin, and hemosiderin; and also in (4) transferrin and cytochrome c (iron-transport proteins) and others [32]. Iron exhibits a wide range of applications, which can be attributed to its variable oxidation (ferrous and ferric) and spin states (high spin and low spin); this makes it flexible to alter the structural properties depending upon the nature of ligand present in its environment. Overall, the immense applications of metals in life has inspired researchers to biomimic the fabrication of nanomaterials based on compounds of metals, to get them into the desired shape and dimensions at the molecular level. The identification and manipulation of the specific ionic, covalent, and noncovalent interactions prevailing between biomolecules and the iron oxide based nanomaterials are very crucial in this aspect [33].

Iron and iron oxides are easily available, less toxic, economic, biocompatible, and thermodynamically stable, which makes them a lucrative option to be used as base material for a variety of multidisciplinary applications. Infrared (IR) spectroscopy has been extremely pivotal in the identification, labeling, and monitoring of these interactions to develop novel synthetic protocols for various iron oxide nanomaterials [34—40]. About a couple of decades back, iron oxide based nanomaterials had limited and conventional areas of application such as geology (as minerals), electrochemistry (for corrosion), biology (for iron-proteins and biomineralization), and industries (as pigments, catalysis, magnetic tapes, and ceramic material). The significant alterations in the physicochemical properties in the nanoregime and the precise synthetic control achieved upon the morphology and dimension of the various nanostructures have extended the applications of iron oxide based nanomaterials such as in lithium ion batteries, magnetic resonance imaging, surface engineering, data storage, drug delivery, hyperthermia, cellular therapy, and various other environmental and nanobiotechnological applications [41—44].

### 6.5.1 Biomolecules: Fe-oxides

Researchers have explored synthesis of various phases of iron oxides by using small molecules and biomolecules. Infrared spectroscopy plays a significant role in analyzing the interaction of these molecules with Fe-center. The extent of application of IR spectroscopy for achieving a deeper insight of interactions involved in nanomaterials at the molecular level has been intensive with the instrumental advancements. Moreover, advanced IR spectroscopy has also been extensively explored to study the type and nature of interactions of iron oxides with different biomolecules and their variations having potential biological applications. The characterization of functionalization of $Fe_3O_4$ nanoparticles by using various capping or stabilizing agents such as (1) small organic molecules (acetylcysteine, aminobenzoic acid, dopamine, citric acid, mercaptoundecanoidc acid and vitamin C); (2) adenine; (3) 5′-guanosine monophosphate, (4) nucleic acids and (5) peptide etc., have been discussed. Various organic and biomolecules based on the functional groups available for IR active interactions with iron nanoparticles have been examined and are listed in Tables 6.3 and 6.4.

### 6.5.1.1 Small organic molecules as stabilizing agents

These organic molecules are well known as capping agent/stabilizing agents for surface functionalization of metal nanomaterial. The most promising capping agents (Fig. 6.4) such as acetylcysteine (16), aminobenzoic acid (17), dopamine (18), citric acid (19), mercaptoundecanoic acid (20), ascorbic acid (21), and dehydroascorbic acid (21) have been greatly utilized for the surface functionalization of $Fe_3O_4$ nanoparticles (NPs). These capping agent/stabilizing agents possess common functional groups including $-NH_2$, $-COOH$, $-SH$, and $-C=O$, etc., which are available for strong interactions with iron/iron oxide nanoparticles [34]. FTIR spectroscopy offers valuable information to characterize

**Table 6.3 IR vibration peaks (cm$^{-1}$) of the various organic capping agents.**

| Sr. no. | Capping agent/ stabilizing agents | Vibrational peaks (in cm$^{-1}$) |
|---|---|---|
| 1 | Acetylcysteine (16) | $-N-H$ bending (1603), weak band $-CH_3$ stretching (2971) |
| 2 | Aminobenzoic acid (17) | $-COO-$asymmetric stretching (1509), $-COO-$symmetric stretching (1383) |
| 3 | Dopamine (18) | $-N-H$ bending (1607), $-C=C-$stretching (1583 and 1479), $-CH_2$ scissoring (1465), $-C-O-$stretching of phenolic $-OH$ (1255) |
| 4 | Citric acid (19) | $-OH$ stretching (3421), $-C-OH$ stretching (1063), OH vibration (1594) |
| 5 | Mercaptoundecanoic acid (20) | $-COO-$asymmetric stretching (1519), $-COO-$symmetric stretching (1413), $-CH_2$ asymmetric stretching (2916), $-CH_2$ symmetric stretching (2844) |
| 6 | Ascorbic acid (21) | $C=O$ band (1755), $C=C$ stretching (1656) |
| 7 | Dehydroascorbic acid (22) | $C=O$ peak (1790) |

**Table 6.4 Infrared spectral data of the different samples corresponding to the various functional groups of GMP, β-FeOOH, GMP-β-FeOOH NPs, GMP-β-FeOOH hydrogel.**

| Group/moiety | GMP (cm$^{-1}$) (observed) | β-FeOOH without using template (cm$^{-1}$) | GMP-β-FeOOH NPs (cm$^{-1}$) Fresh (observed) | GMP-β-FeOOH colloidal hydrogel (cm$^{-1}$) aged (observed) |
|---|---|---|---|---|
| $>C(6)=O$ | 1696 (s) | — | 1676 (w & sh) | 1676 (w & sh) |
| $-NH_2$ | 1653 (sh) | — | 1637 (s) | 1637 (s) |
| C=N and ring skeletal vibrations | 1607 (m) | | 1600 (sh) | 1600 (sh) |
| Pyrimidine/ Imidazole vibration | 1535 (s) | — | 1535 (sh) | Almost disappeared |
| N(7)-C(8) stretching, C(8)-H Bending | 1481 (s) | — | 1481(m) | 1482 (w) |
| Imidazole | 1416 (m) | | 1409 (w) | 1402 (w) |
| Imidazole | 1371 (m) | — | 1357 (w) | 1359 (sh) |

(Continued)

**Table 6.4 (Continued)**

| Group/moiety | GMP (cm$^{-1}$) (observed) | $\beta$-FeOOH without using template (cm$^{-1}$) | GMP-$\beta$-FeOOH NPs (cm$^{-1}$) Fresh (observed) | GMP-$\beta$-FeOOH colloidal hydrogel (cm$^{-1}$) aged (observed) |
|---|---|---|---|---|
| Pyrimidine | 1256 (br) | — | 1261 (w) | 1259 (w) |
| $\upsilon$–C–C (sugar) | 1180 | | — | — |
| $\upsilon$–C–O (sugar) | 1113 (sh) | | 1109 (sh) | 1108 (sh) |
| $PO_3^{-2}$ antisymmetric stretching | 1090 (br) | — | 1069 (br) | 1084 (br) (shape is changed) |
| $PO_3^{-2}$ symmetric stretching | 978 (s) | — | 991 (m) | 989 (sh) |
| Sugar ring | 905 (w) | | 904 (w) | — |
| Sugar ring | 866 (w) | — | 868 (w) | — |
| P–O–5'-sugar C2'-endo/anti conformer | 806 (m) | — | 799 (w) | 802 (almost disappeared) |
| P–O | 780 (m) | — | 782 (sh) | 781 (w) |
| Ring mode | 625 (w) | — | 635 (w) | — |
| Skeletal deformation | 535 (w) | — | — | — |
| $H_2O$ bending | — | 1634 (s) | 1637 (s) | 1637 (s) |
| O–H–Cl deformation | — | 833, | — | — |
| Fe–O–Fe stretching | — | 696, 644, 471, 420 | 681(br), 635 (w), 498 (sh), 483, 472 | 687 (br), 631 (w), 470 (br) |
| Additional peaks | 1241, 724, 692, 580 | — | 1383, 799, 606 | 1460, 1402, 1018 |

S, sharp; m, medium; w, weak; br, broad; sh, shoulder.

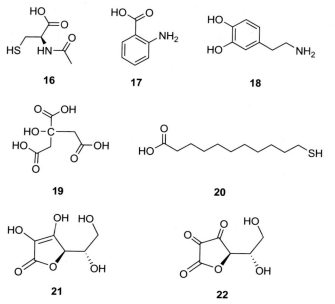

Figure 6.4 Molecular structure different organic capping agents: acetylcysteine (**16**), aminobenzoic acid (**17**), dopamine (**18**), citric acid (**19**), mercaptoundecanoic acid (**20**), ascorbic acid (**21**), and dehydroascorbic acid (**22**).

the detailed surface composition of stabilizing agent-functionalized nanoparticles by analyzing interactions between the surfaces of iron nanoparticles and capping agents. The IR peaks observed in spectra with reference to surface change before and after functionalization of nanoparticles with stabilizing agent reveals significant structural changes on the surface of nanoparticles (Fig. 6.5). The spectrum originates from $Fe_3O_4$ nanoparticles without stabilizing agents showed a broad and strong band at 3421 and 1063 cm$^{-1}$ assigned to O−H and C−OH stretching modes, respectively. The IR band at 1594 cm$^{-1}$ corresponding to $\delta$OH vibrations originates from hydration and hydrogen bond formation. The weak IR bands at 2921, 2853, and 1454 cm$^{-1}$ were allotted to symmetric, asymmetric stretching, and scissoring vibration of $CH_2$ from diethylene glycol (DEG), respectively. Further, acetylcysteine (16), aminobenzoic acid (17), and citric acid (19) coated magnetic nanoparticles endowed strong characteristic band due to COO−stretching, this stretching peak is absent in case of DEG.

The acetylcysteine functionalized nanoparticles showed N−H bending peak at 1603 cm$^{-1}$ whereas in case of aminobenzoic acid, peaks at 1605, 3367, and 1302 cm$^{-1}$ were assigned to N−H bending, N−H stretching, and aromatic C−N stretching [34]. Moreover, dopamine (18) functionalized $Fe_3O_4$ exhibited characteristic peak of catechol. The IR band at 1607 cm$^{-1}$ correlated to N−H bending and 1583 and 1479 cm$^{-1}$ attributed to C=C stretching, however intense band at 1255 cm$^{-1}$ was aroused due to C−O stretching. The mercaptoundecanoic (20) acid as stabilizing agent with $Fe_3O_4$ displayed strong band at 1519 and 1413 cm$^{-1}$ due to asymmetric COO−stretching and symmetric stretching of COO−, respectively. Due to the long aliphatic chain present in

mercaptoundecanoic acid, intense asymmetric and symmetric bands appear at 2916 and 2844 cm$^{-1}$ by $CH_2$ group stretching [34]. Overall the characteristic peaks/bands used in the analysis and characterization of surface composition of $Fe_3O_4$ NPs functionalized with stabilizing agents are listed in Table 6.3.

Furthermore, IR spectroscopy has also been used in predicting the oxidized and reduced form of biomolecules and also to predict whether any of these forms have been chemically functionalized on the surface of iron oxide NPs ($Fe_3O_4$). Xiao et al. reported the comparative FTIR spectroscopic analysis to study the chemical transformation of ascorbic acid (21, oxidized form) to dehydroascorbic acid (22, DHAA, reduced form) and also examine the surface functionalization of iron oxide NPs with DHAA (Fig. 6.6) [35]. In fact, the absence of C=O stretching bands in DHAA-$Fe_3O_4$ NPs and emergence of new absorption frequency at 1620 cm$^{-1}$ reflected the coordination of Fe-center to the surface of $Fe_3O_4$ nanoparticles through oxygen atom of the carbonyl group.

### 6.5.1.2 Nucleic bases—β-FeOOH nanoparticles

FTIR spectroscopy has played vital role in unraveling the mechanism of nucleation and growth of supermagnetic β-FeOOH nanostructures in presence of nucleic bases. Adenine (23) (Fig. 6.7) based β-FeOOH nanoparticles have been fabricated by hydrolysis of Fe(III) chloride by employing varying concentrations of adenine [36].

The FTIR analysis revealed the iron oxide (β-FeOOH) interactions with adenine, primarily through −$NH_2$, N(3) of the pyrimidine ring and N(7) and N(9)H of imidazole

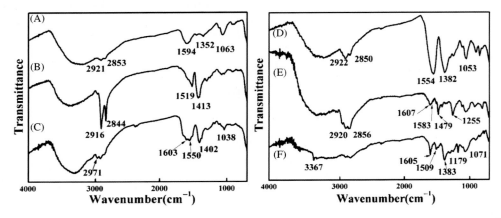

**Figure 6.5** FTIR spectra of $Fe_3O_4$ nanoparticles: (A) without stabilizing agent and (B) with stabilizing agent mercaptoundecanoic acid, (C) acetylcysteine, (D) citric acid, (E) dopamine, (F) aminobenzoic acid [34].

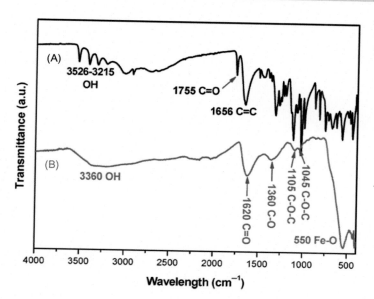

Figure 6.6 FTIR spectra of the ascorbic acid or Vitamin C (A) and the as-prepared iron oxide NPs (B) [35].

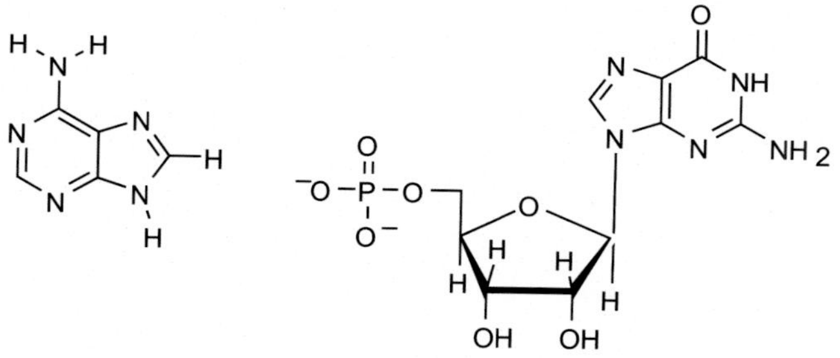

Figure 6.7 Chemical structures of adenine (23) and 5′-Guanosine monophosphate (24).

23

24

ring. The interactions developed as a result of capping β-FeOOH nanoparticles with adenine that induced the morphological transition from solid nanorods to a mixture of porous nanorods and spherical NPs by subsequently increasing the amount of adenine [36]. Interactions of β-FeOOH nanostructures with adenine also enhanced the solubility and colloidal stability of the β-FeOOH nanostructures. In the IR spectrum (Fig. 6.8) the vibration bands at 815 and 693 cm$^{-1}$ corresponding to Fe—OH—Cl and Fe—O stretching might have resulted due to the deformation, while IR peak at 422 cm$^{-1}$ and a shoulder at 467 cm$^{-1}$ is assigned to Fe—O—Fe stretch [36]. The significant decrease in the intensity of IR absorption peak at 819 cm$^{-1}$ is due to Fe—OH—Cl deformation, which might be responsible for rod-like shape formation upon increasing adenine concentration. These interactions

provide valuable information to understand this morphological transition.

### 6.5.1.3 5′-Guanosine monophosphate-β-FeOOH

IR spectroscopy is not only useful in identification of simple interactions but it also helps in the investigation of the extensive supramolecular interactions between 5′-Guanosine monophosphate (GMP, 24) (Fig. 6.9) templated β-FeOOH colloidal nanostructures. The aging of these nanostructures for a period of 1 week transformed the colloidal solution into porous GMP-β-FeOOH hydrogel at room temperature [37]. The GMP-stabilized colloidal β-FeOOH nanostructures were prepared by the hydrolysis of Fe (III) chloride solution in varying

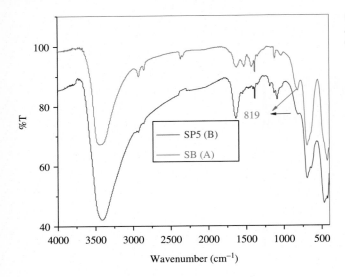

**Figure 6.8** FTIR spectra of β-FeOOH nanostructures (A) in absence of adenine (*red color*), (B) in presence of adenine (*blue*) [36].

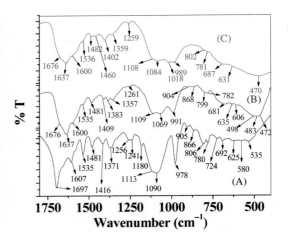

**Figure 6.9** FTIR spectra of (A) pure GMP (*black*), (B) GMP-β-FeOOH colloidal solution (*blue*), and (C) GMP-β-FeOOH hydrogel (*red*) [37].

concentration of GMP solution. Interestingly, it was observed that in the presence of metal ion, the supramolecular interactions were found to be so intensified that it resulted in the gellification with GMP-β-FeOOH colloidal solution even with $10^3$ times lesser concentration of GMP used as compared with the reports available for pure GMP-hydrogels [45−48].

The IR spectra of GMP-stabilized hydrogel along with that of pure β-FeOOH and pure GMP processed under identical synthetic conditions are depicted in Fig. 6.9 and respective data of fabricated nanostructures is compiled in Table 6.4. A comparative analysis of the FTIR spectrum of GMP and GMP-β-FeOOH colloidal solution samples

indicates that the vibrational peaks and nature of these bands of GMP-β-FeOOH are fairly different from pure GMP. The spectral data indicated that GMP interacts with β-FeOOH in GMP-β-FeOOH colloidal solution through various moieties and functional groups such as sugar ring, imidazole, pyrimidine, −C=O, −NH$_2$, −C=N−present in GMP, −C−O−of the sugar, −P−O−5′-sugar, PO$_3^{-2}$ and −P−O (Table 6.4) [37].

The IR spectrum of GMP-β-FeOOH hydrogel shows additional deviations in the position as well as in the shapes of peak due to enhanced supramolecular interactions that emerged from −C=O, −NH$_2$, −P−O−5′-sugar, PO$_3^{-2}$, and Fe−O−Fe. The majority of these peaks have been broadened as compared with the fresh colloidal GMP-β-FeOOH sample (Table 6.4). The interactions of β-FeOOH were observed specifically through the pyrimidine and/or imidazole ring of GMP, $-PO_3^{-2}$ and sugar ring of GMP. Moreover, vibrational peaks appeared due to the original pyrimidine and/or imidazole ring and sugar moieties disappearing. A significant change was noted in the vibrational frequencies and shape of bands for symmetric and antisymmetric stretching of $-PO_3^{-2}$ [37]. The variations in the nature of vibrational peaks/bands indicated the reorganization of the interactions of GMP template with β-FeOOH resulting in gellification.

### 6.5.1.4 DNA- Fe (II) and Fe (III) nanoparticles

Nucleic acids, particularly deoxyribonucleic acid (DNA), have been extensively explored in fabrication of nanomaterial due to their biocompatibility and various polar functional groups. Ouameur et al. reported the possible interactions of DNA with Fe (II) and Fe (III) using IR spectroscopy [39]. IR vibrational peaks have been used to distinguish and define the functionalities/moieties present in DNA and their interaction with the ferrous and ferric centers. During the study of sodium salt of calf thymus DNA solution with varying concentrations of ferrous and ferric salts in different concentration ratio, that is, Fe: DNA = 1:160; 1:80; 1:40; 1:20; 1:10; 1:4; and 1:2, confirmed shift from the major IR peaks observed in frequency range 1717–1708 $cm^{-1}$. Furthermore, IR spectra depicted a major spectral shift of bands 1717–1708 $cm^{-1}$ and peak at 1222–1218 $cm^{-1}$ at lower concentrations of Fe and DNA ratio (Fe:DNA = 1:80–1:40). The characteristic vibrational frequency at 1708 $cm^{-1}$ was assigned to chelation Fe (II) with N atom at the 7th position in guanine (G) and peak at 1218 $cm^{-1}$ arose due to asymmetric stretching of $PO_2$. The ratios of the intensity of $PO_2$ symmetric stretching (1088 $cm^{-1}$) and asymmetric stretching (1222 $cm^{-1}$) were also observed to change upon complexation of Fe(II) with DNA with $\nu_s/\nu_{as}$ decreasing from 1.75 to 1.55 upon complexation. In case of thymine (T) and adenine (A), only increase in the intensity of peaks at 1663 (T) and 1609 $cm^{-1}$ (A) was observed, while no significant spectral shift [39] was noticed. Upon increasing the concentration of ferrous ion (Fe:DNA > 1:10), DNA-in plane vibrations for nitrogen bases; only guanine (1717 $cm^{-1}$), thymine (1663 $cm^{-1}$), adenine (1609 $cm^{-1}$) and $PO_2$ (1222 $cm^{-1}$) were majorly observed to increase in intensity. However, the IR spectra endowed no significant interaction of Fe (II) with cysteine bases. Concisely, it has been confirmed that Fe (II) interacts with N atom in the 7th position of guanine in DNA and also with backbone $PO_2$ group [39].

Although at lower concentration of ferric ions (Fe: DNA = 1:80), Fe (III) was observed to bind with the backbone $PO_2$ group without any agitation of the nitrogen bases, which was revealed from a minor increment in the intensity of stretching mode of phosphate detected at 1222 $cm^{-1}$. However, a major reduction in the intensity of frequency at 1717 (G), 1663 (T), and 1609 $cm^{-1}$ (A) might be attributed to the helix stability due to the chelation of Fe-phosphate. The trend of variations observed in the intensity of $PO_2$ symmetric and asymmetric vibrations with respect to change in ratio of Fe: DNA was quite similar to that observed in the case of Fe (II)-DNA complex. The IR spectra recorded upon increasing the concentrations (Fe: DNA = 1:40−1:20) depicted similar interactions as observed for complexation of Fe (II) and DNA, that is, with N atom in the 7th position of guanine and $PO_2$ backbone, but the peak at 1717 $cm^{-1}$ shifted to 1712 $cm^{-1}$ with a 30% increase in its intensity and band 1222 $cm^{-1}$ with 20% enhancement in intensity [39]. In contrast, no such increase in the intensity or shift in the position of the bands was observed for thymine and adenine moieties.

### 6.5.1.5 Peptide mediated biomineralization of Iron (III) oxyhydroxide nanoparticles

FTIR spectroscopy has also been employed for the mechanistic study of the biological formation of monodispersed iron(III) oxyhydroxide NPs functionalized by peptides in muscle protein hydrolysate (AMPH) obtained from anchovy (*Engraulis japonicas*) [38]. The FTIR spectroscopic analysis was specifically used to analyze the effect of pH on the formation of peptides mediated by Fe-oxyhydroxide nanoparticles. The peptide scaffold bonded with Fe through carboxyl group showed 27.5 mg iron $g^{-1}$ peptide iron-loading capacity. FTIR spectra of apo myosin and Fe-loaded myosin at different pH revealed that at higher pH (pH = 8.0), no significant differences in the two spectra were observed. However, at lower pH such as 1.0, 3.0, and 5.0, the IR data clearly indicated a decline in the intensity of 1716 $cm^{-1}$ peak (carboxylic group of myosin) and an increase in the peak intensity at 3420 $cm^{-1}$ as compared with apo forms. This increase in the absorption intensity was allotted to the H−OH stretching of water molecules adsorbed by Fe-loaded myosin.

## 6.5.2 Guanosine monophosphate-cadmium sulfide nanostructures

Kumar et al. employed FTIR to analyze the interactions responsible for the formation of guanosine monophosphate

(GMP) mediated cadmium sulfide (CdS) nanostructures. To investigate the mode of interactions of CdS nanostructures with GMP, FTIR spectra including sodium salt of GMP (Na$_2$–GMP), GMP–Cd$^{2+}$, and GMP–CdS were examined under identical experimental conditions as shown in Fig. 6.10A–C, respectively [5]. As Fig. 6.10A depicts, the IR spectrum of Na$_2$–GMP yields characteristic peaks at 1694 and 1640 cm$^{-1}$, assigned to >C=O and –NH$_2$ group, respectively. Sharp band at 1363 and 1236 cm$^{-1}$ has been assigned to imidazole and pyrimidine moiety. IR peaks at 1080 and 822 cm$^{-1}$ were due to phosphate $-(PO_3^{-2})$ and P–O–5'-sugar respectively.

As shown in Table 6.5, a shift in frequencies from 1694, 1640, 1490, 1363, 1236, 1080, and 822 cm$^{-1}$ in Na$_2$-GMP to 1691, 1638, 1471, 1088, and 806 cm$^{-1}$ upon binding with Cd$^{2+}$, clearly indicates the binding of cadmium ion (Cd$^{2+}$) to GMP though various functional groups including carbonyl (>C=O), amine (–NH$_2$), phosphate $(-PO_3^{-2})$, and P–O–5'- of sugar that are easily available for interactions (Table 6.5) [5]. The formation of GMP functionalized CdS nanoparticles induced further shift in frequencies assigned to the functional groups. Moreover, nature and shape of IR peaks assigned to the five-membered imidazole ring and six-membered pyrimidine ring of GMP and their respective vibrational shifts were observed and tabulated [5]. In Cd$^{2+}$-GMP structures, the interaction of Cd$^{2+}$ to GMP induces the enhancement in the intensity of peaks arising due to the antisymmetric stretching of PO$_3^{-2}$ and a slight reduction in the intensity of P–O stretching due

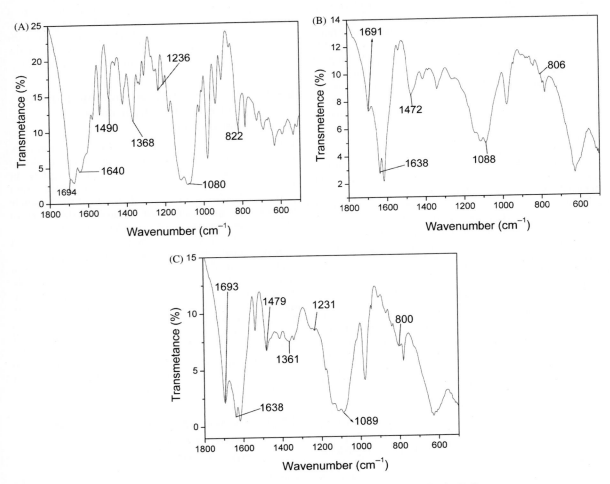

Figure 6.10 (A) FTIR spectra of Na$_2$-GMP; (B) FTIR spectra of Cd$^{2+}$-GMP; (C) FTIR spectra of CdS-GMP.

**Table 6.5** Significant peaks observed in IR spectra of Na$_2$-GMP, Cd$^{2+}$-GMP, and CdS-GMP.

| Group/moiety | Na$_2$-GMP (cm$^{-1}$) | Cd$^{2+}$-GMP (cm$^{-1}$) | CdS-GMP (cm$^{-1}$) |
|---|---|---|---|
| >C=O | 1694 (sh) | 1691 (m) | 1693 (s) |
| —NH$_2$ | 1640 (w) | 1638 (s) | 1638 (s) |
| N$^7$—C$^8$ + C$^8$—H | 1490 (s) | 1471 (m) | 1479 (m) |
| Imidazole | 1363 (s) | — | 1361 (br) |
| Pyrimidine | 1236 (s) | — | 1231 (sh) |
| PO$_3^{-2}$ | 1080 (m) | 1088 (m) | 1089 (br) |
| P—O—5'-sugar | 822 (s) | 806 (sh) | 800 (br) |

to P—O—5'-sugar was observed as shown in Fig. 6.10B and Table 6.5. In contrast, the CdS- GMP nanohybrid formation induces the minor decrease in frequency of P—O stretching due to P—O—5'-sugar; however, no appreciable change was noticed in the frequency allotted to antisymmetric stretching of PO$_3^{-2}$ (Table 6.5). Concisely, this study provided the confirmation that the cadmium ion (Cd$^{2+}$) largely interacts via the anionic oxygen of phosphate (PO$_3^{-2}$) while cadmium sulfide (CdS) prefers to bind with ethereal oxygen of P—O—5'-sugar.

## 6.6 Graphene-based nanomaterials

Graphene, the mother of all graphitic materials, has attracted immense attention over the past several years in various research areas, mainly electronics, energy storage and conversion, and photonics due to its interesting physicochemical properties [49−51]. Specifically, in the biomedical research field, graphene has shown immense potential owing to its extraordinary mechanical strength, transparency, electrical conductivity, and also the biocompatibility [52−54]. Particularly, the unique structure of graphene oxide (GO) consists of $sp^2$ and $sp^3$-hybridized carbon atoms responsible for hydrophobic property and presence of abundant oxygen functionalities such as hydroxyl (—OH) and epoxy groups on its basal plane offers hydrophilic nature, and carboxyl groups on its edges make the GO sheets amphiphilic in nature [53−55]. With the aid of these hydrophilic oxygenated functionalities, GO can be dispersible in many polar solvents as well as in aqueous solvents. The π-electrons clouds of $sp^2$-hybridized carbon at the basal plane of GO are able to form π−π interactions with the aromatic moieties of various scaffolds.

The characterization of structure and functionality of GO has been performed by various analytical techniques but FTIR plays an important role. FTIR is simple, quick, and the most efficient technique for determining the various functional groups present on the surface of GO and residual functional groups in reduced GO (rGO). Fig. 6.11 shows the typical FTIR spectrum of GO synthesized by the oxidation of graphite following Hummers' methods in the frequency region of 4000−500 cm$^{-1}$ [56]. The intense vibrational peak at 3425 cm$^{-1}$ corresponds to the O—H stretching from water molecules. The two absorption peaks in IR spectrum at 1718 and 1632 cm$^{-1}$ were allotted to the stretching modes of —C=O of —COOH and —C=C—, respectively. The other strong and intense vibrational bands at 1373, 1222, and 1054 cm$^{-1}$ have been attributed to the bending mode of C—O—C (epoxy), tertiary C—OH, and C—O (alkoxy) groups, respectively (Fig. 6.11) [56,57]. However, the position of these characteristic peaks corresponding to different functional groups may vary slightly from work to work due to the difference in experimental conditions.

Until now, a variety of materials have been employed for functionalization of GO such as chitosan, dextran, collagen, folic acid, polyethylene glycol, poly-L-lysine, polyethylenimine, polyacrylic acid, poly(vinyl alcohol), bovine serum albumin, DNA, RNA, amino groups, sulfonic groups, metal NPs (Au, Pt, and Ag), and metal oxide NPs (iron oxide) exhibiting the covalent/noncovalent interactions to increase the aqueous dispersibility, stability in physiological solution, and to minimize the toxicity [52−54,58,59].

Recently, Liu et al. [60] have demonstrated the surface chemistry driven approach for switching on/off the interaction between GO and doxorubicin (DOX) with the aim of development of drug delivery system for DOX drug by loading and releasing of drug on GO. The IR spectrum of GO-DOX showed the noncovalent bonding between GO and DOX (Fig. 6.12) and exhibited the absorption bands assigned to both the GO and DOX, which indicated the binding of DOX on GO. The typical FTIR spectrum of DOX endowed vibrational band at 1615 cm$^{-1}$ corresponding to C=C stretching, which gets blue shifted to 1620 cm$^{-1}$ after loading on GO (GO-DOX). Furthermore, the peak corresponding to the stretching mode of C=O

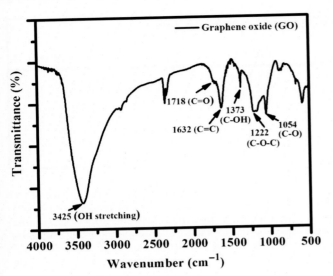

Figure 6.11 FTIR spectrum of graphene oxide (GO) [56].

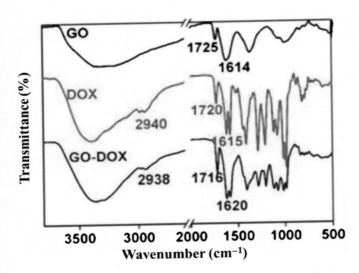

Figure 6.12 FTIR spectra of GO (*black*), DOX (*red*), and GO-DOX (*blue*) demonstrating the noncovalent interaction between GO and DOX [60].

assigned at frequency 1720 cm$^{-1}$ in DOX gets red shifted to 1716 cm$^{-1}$ after GO loading [60]. The shift in stretching vibration frequencies in the spectrum of GO-DOX compared with GO and DOX clearly indicates that this shift might be due to the electron transfer mechanism between DOX and GO.

Kolanthai et al. [61] reported the construction of novel biodegradable composites composed of basic scaffold alginate-chitosan-collagen (SA-CS-Col) and incorporation with GO to enhance the porous assembly to offer prospective applications specifically in bone tissue engineering. The structural property of the synthesized biodegradable

SA-CS-Col-GO scaffolds was studied by FTIR spectroscopy, which revealed the bond formation in SA-CS-Col scaffolds with GO via ionic interactions. Further, the FTIR studies also disclosed that the addition of GO to the basic SA-CS-Col scaffolds increased the intensity of −OH group in the SA-CS-Col-GO scaffolds due to intermolecular H−bonding. This bonding promoted the interfacial adhesion and also enhanced the mechanical properties of the synthesized biodegradable scaffolds that might be useful for tissue engineering. This synthesized composites were found to be more stable than basic scaffold without GO. [61] For instance, Kumar and Khandelwal [56]

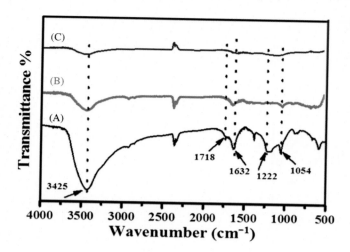

**Figure 6.13** IR spectra of (A) graphene oxide (black), (B) reduced graphene oxide (rGO) formed by reduction of malonic acid (red), (C) As-synthesized rGO annealed at 300°C (blue) [56].

demonstrated the reduction of GO employing reducing agent malonic acid at 95°C for 6 hours under basic pH conditions (10.5) to construct the reduced form of GO resulting into rGO sheets. The extent of GO reduction was analyzed by FTIR spectroscopy, revealing the significant decrease in intensity of vibrational peaks due to the stretching of hydroxyl (−OH) and alkoxy (C−O) groups and almost complete removal of the peaks ascribed to C=O (COOH) and C−O−C (epoxy) (Fig. 6.13).

The IR spectra (A–C) shown in Fig. 6.13 clearly indicate the peak intensity at 3425 assigned for −OH, 1718 allocated for −C=O, 1632 for C=C, 1222 for epoxy, and 1054 cm$^{-1}$ for C−O (alkoxy) functional group present in graphene oxide. While in IR spectrum of reduced graphene oxide formed by reduction of malonic acid, the IR signal at 3425 cm$^{-1}$ arose due to −OH stretching and peak at 1054 cm$^{-1}$ allocated to C−O was significantly reduced while peaks at 1718 and 1222 cm$^{-1}$ assigned to −C=O and −C−O−C−completely disappeared. [56] This data confirmed the fact that $sp^2$ character has increased in the reduced graphene oxide compared with graphene oxide. Moreover, in case of reduced graphene oxide annealed at 300°C, shows further reduction in the intensity of IR absorption peak at 3425 and 1054 cm$^{-1}$ assigned to −OH stretching and C−O functional moieties that indicates the elimination of oxygen moieties from graphene oxide [56]. Further, the same research group reported the formation of graphene nanoribbons and their characterization by IR spectroscopy [62]. The formation of these graphene nanoribbons has been explained by the supramolecular interactions that were observed in IR spectrum, between −COOH groups present on graphene oxide and with malonic acid, as displayed in Fig. 6.14.

Several reviews have focused on the advancements in graphene-based nanomaterials for biotechnological and biomedical functions [63,64] in recent years. In 2015, Kumar et al. [65] published the composite of poly (ε-caprolactone) with GO, rGO, and amine functionalized graphene oxide (AGO). The reduction and functionalization of GO with amine group was ascertained by FTIR measurements as shown in Fig. 6.15. The FTIR spectrum of AGO is quite different to that of GO showing the N−H stretching peaks as doublet by exhibiting absorbance band at ~ 3368 and 3215 cm$^{-1}$. In addition to this, some important additional peaks have also been marked at vibrational frequency at 1512, 1260, and 800 cm$^{-1}$ corresponding to phenyl group of methylenedianiline (MDA), stretching of −C−N in aryl group, and N−H bending of NH functional group, respectively [65]. The fabricated functionalized graphene in polymer composites exhibited enhanced mechanical and biological properties. Hitherto, a number of biomolecules, polymers, and metal/metal oxide NPs have been employed for the modification/functionalization of rGO/graphene for a wide range of biological applications [52−54].

In recent times, zero-dimensional graphene GQDs, described as graphene sheets having dimensions less than 100 nm and the thickness of less than 10 layers, have also have gained immense attention owing to their small size, tunable photoluminescing properties, high photostability, biocompatibility, chemical inertness, ease of functionalization, and interesting physicochemical properties. These features make them interesting candidates for various biological applications in bioimaging, biosensing, drug/gene delivery, and other theranostics [66,67].

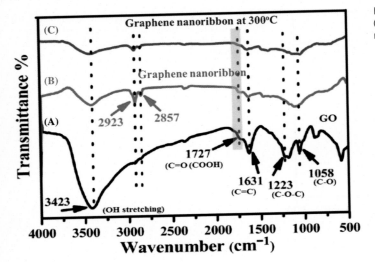

Figure 6.14 FTIR spectra of (A) Graphene oxide, (B) graphene nanoribbon, and (C) graphene nanoribbons at 300°C [62].

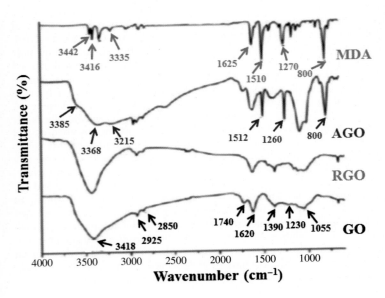

Figure 6.15 FTIR spectra of GO, rGO, and AGO [65].

## 6.7 Conclusion

We have discussed the application of IR spectroscopy in structural elucidation of organic molecules, analysis of surface composition, distinguishing the binding molecular species, nucleation and growth of nanostructures, gellification, biomineralization and interaction of biomolecules with different oxidation states of Fe. On the premise of above reports, IR spectroscopy has certainly proven to be an indispensable analytical technique for providing precise and cutting edge information in the characterization of the multiple aspects of bioinorganic interactions for biomolecule–inorganic nanosystems with promising future applications. In light of these motives, a few advancements have already been made by coupling IR spectroscopy with other characterization techniques such as atomic force microscopy and gas chromatography for performing real-time and high-speed nanometric analysis. The developments in spectroscopic instrumentation in last decade have unfolded a vast realm for the applications for IR spectroscopy for its future applications in nanobiotechnology. Near IR, time-resolved IR, and 3D-FTIR spectroscopic studies have enough ground to be employed in the area of various biosystems for the in vitro and in vivo studies, which will certainly help in biomimicking and artificial biomineralization of organic and inorganic nanostructures with enhanced physicochemical properties.

| | |
|---|---|
| GO | graphene oxide |
| GO-DOX | graphene oxide functionalized doxorubicin |
| GQDs | graphene quantum dots |
| IR | infrared |
| NPs | nanoparticles |
| PEP | phosphoenolpyruvate |
| PK | pyruvate kinase |
| rGO | reduced GO |
| SA-CS-Col | alginate-chitosan-collagen scaffolds |

## 6.8 List of abbreviations

| | |
|---|---|
| ADP | adenosine diphosphate |
| AGO | amine functionalized graphene oxide |
| DEG | diethylene glycol |
| DHAA | dehydroascorbic acid |
| DNA | deoxyribonucleic acid |
| DOX | doxorubicin |
| FTIR | Fourier transform infrared |
| GMP | 5′-guanosine monophosphate |

## Acknowledgments

RR and VK are thankful to Amity University Noida for technical support and DST-SERB for financial assistance. AK is thankful to Amity University Greater Noida. MK and SK Gupta are thankful to University of Ulsan South Korea and Harcourt Butler Technical University, Kanpur.

**Conflict of interest statement**
The author(s) have no conflicts of interest.

## References

[1] Dominguez G, Mcleod AS, Gainsforth Z, Kelly P, Bechtel HA, Keilmann F, et al. Nanoscale infrared spectroscopy as a non destructive probe of extraterrestrial sample. Nat Commun 2014;5(5445):1−10. Available from: https://doi.org/10.1038/ncomms6445.

[2] Chaber R, Łach K, Szmuc K, Michalak E, Raciborska A, Mazur D, et al. Application of infrared spectroscopy in the identification of Ewing sarcoma: a preliminary report. Infrared Phys Technol 2017;83:200−5. Available from: https://doi.org/10.1016/j.infrared.2017.05.006.

[3] Diem M. Introduction to modern vibrational spectroscopy. New York, NY: Wiley; 1993. Available from: https://doi.org/10.1002/bbpc.19940981029.

[4] Kumar A, Kumar V. Supramolecular-directed synthesis of RNA-mediated CdS/ZnS. Nanotubes. Chem Commun 2009;0(36):5433−5. Available from: https://doi.org/10.1039/B907283G.

[5] Kumar A, Kumar V. Synthesis and optical properties of guanosine 5-monophosphate-mediated CdS nanostructures: an analysis of their structure, morphology, and electronic properties. Inorg Chem 2009;48(23):11032−7. Available from: https://doi.org/10.1021/ic901205c.

[6] Kumar A, Kumar V. Self-assemblies from RNA-templated colloidal CdS nanostructures. J Phys Chem C 2008;112(10):3633−40. Available from: https://doi.org/10.1021/jp7109803.

[7] Sondhi SM, Rani R, Gupta PP, Agrawal SK, Saxena AK. Synthesis, anticancer, and anti-inflammatory activity evaluation of methane sulfonamide and amidine derivatives of 3,4-diaryl-2-imino-4-thiazolines. Mol Divers 2009;13 (3):357−66. Available from: https://doi.org/10.1007/s11030-009-9125-0.

[8] Sondhi SM, Rani R, Diwvedi AD, Roy P. Synthesis of some heterocyclic imides and azomethine derivatives under solvent free condition and their anti-inflammatory activity evaluation. J Heterocyclic Chem 2009;46 (6):1369−74. Available from: https://doi.org/10.1002/jhet.249.

[9] Sondhi SM, Rani R. A convenient, solvent free and high yielding synthesis of bicyclo-heterocyclic compounds. Lett Org Chem 2008;5(1):51−4. Available from: https://doi.org/10.2174/157017808783330180.

[10] Ellis DI, Goodacre R. Metabolic fingerprinting in disease diagnosis: biomedical applications of infrared and Raman spectroscopy. Analyst 2006;131(8):875−85. Available from: https://doi.org/10.1039/b602376m.

[11] Kumar A, Kaloti M, Navani NK. Synthesis of glucose-mediated Ag-γ-Fe2O3 multifunctional nanocomposites in aqueous medium - a kinetic analysis of their catalytic

activity for 4-nitrophenol reduction. Green Chem 2015;17(10):4786—99. Available from: https://doi.org/10.1039/C5GC00941C.

[12] Bayda S, Hadla M, Kumar V, Palazzolo S, Ambrosi E, Pontoglio E, et al. A bottom-up synthesis of carbon nanoparticles with better doxorubicin efficacy. J Control Release 2017;248:144—52. Available from: https://doi.org/10.1016/j.jconrel.2017.01.022.

[13] Sondhi SM, Rani R, Roy P, Agrawal SK, Saxena AK. Microwave-assisted synthesis of N-substituted cyclic imides and their evaluation for anticancer and anti-inflammatory activities. Bioorg Med Chem Lett 2009;19(5):1534—8. Available from: https://doi.org/10.1016/j.bmcl.2008.07.048.

[14] Rani R, Arya S, Kilaru P, Kumar N, Roy P, Sondhi SM. An expeditious, highly efficient, catalyst and solvent-free synthesis of 9,10-dihydro-anthracene-9,10-$\alpha,\beta$-succinimide derivatives. Green Chem Lett Rev 2012;5(4):545—75. Available from: https://doi.org/10.1080/17518253.2012.677069.

[15] Arya S, Kumar S, Rani R, Sondhi SM. Synthesis, anti-inflammatory, and cytotoxicity evaluation of 9,10-dihydroanthracene-9,10-$\alpha,\beta$-succinimide and bis-succinimide derivatives. Med Chem Res 2013;22(9):4278—85. Available from: https://doi.org/10.1007/s00044-012-0439-6.

[16] Sondhi SM, Rani R, Roy P, Agrawal SK, Saxena AK. Conventional and microwave assisted synthesis of small molecule based biologically active heterocyclic amidine derivatives. Eur J Med Chem 2010;45(3):902—8. Available from: https://doi.org/10.1016/j.ejmech.2009.11.030.

[17] Sondhi SM, Rani R, Roy P, Agrawal SK, Saxena AK. Synthesis, anti-inflammatory & anticancer activity evaluation of some heterocyclic amidine and bis-amidine derivatives. J Heterocycl Chem 2011;48(4):921—6. Available from: https://doi.org/10.1002/jhet.658.

[18] Sondhi SM, Singh J, Rani R, Gupta PP, Agrawal SK, Saxena AK. Synthesis, anti- inflammatory and anticancer activity evaluation of some novel acridine derivatives. Eur J Med Chem 2010;45(2):555—63. Available from: https://doi.org/10.1016/j.ejmech.2009.10.042.

[19] Sondhi SM, Kumar S, Rani R, Chakraborty AK, Roy P. Synthesis of bis-acridine derivatives exhibiting anticancer and anti-inflammatory activity. J Heterocycl Chem 2013;50(3):252—60. Available from: https://doi.org/10.1002/jhet.985.

[20] Sondhi SM, Rani R. Microwave-mediated one-step synthesis of tri and tetracyclic heterocyclic molecules. Green Chem Lett Rev 2010;3(2):115—20. Available from: https://doi.org/10.1080/17518250903583706.

[21] Sondhi SM, Rni R, Singh J, Roy P, Agrawal SK, Saxena AK. Solvent free synthesis, anti-inflammatory & anticancer activity evaluation of tricyclic and tetracyclic benzimidazole derivatives. Bioorg Med Chem Lett 2010;20(7):2306—10. Available from: https://doi.org/10.1016/j.bmcl.2010.01.147.

[22] Zscherp C, Barth A. Reaction-induced infrared difference spectroscopy for the study of protein reaction mechanisms. Biochemistry 2001;40(7):1875—82. Available from: https://doi.org/10.1021/bi002567y.

[23] Barth A, Zscherp C. Substrate binding and enzyme function investigated by infrared spectroscopy. FEBS Lett 2000;477(3):151—6. Available from: https://doi.org/10.1016/S0014-5793(00)01782-8.

[24] Carey PR, Tonge PJ. Unlocking the secrets of enzyme power using Raman spectroscopy. Acc Chem Res 1995;28(1):8—13. Available from: https://doi.org/10.1021/ar00049a002.

[25] Kumar S, Barth A. Phosphoenolpyruvate and $Mg^{2+}$ binding to pyruvate kinase monitored by infrared spectroscopy. Biophys J 2010;98(9):1931—40. Available from: https://doi.org/10.1016/j.bpj.2009.12.4335.

[26] Gasper R, Mijatovic T, Bénard A, Derenne A, Kiss A, Goormaghtigh E. FTIR spectral signature of the effect of cardiotonic steroids with antitumoral properties on a prostate cancer cell line. Biochim Biophys Acta - Mol Basis Dis 2010;1802(11):1087—94. Available from: https://doi.org/10.1016/j.bbadis.2010.07.012.

[27] Derenne A, Van Hemelryck V, Lamoral-Theys D, Kiss R, Goormaghtigh E. FTIR spectroscopy: a new valuable tool to classify the effects of polyphenolic compounds on cancer cells. Biochim Biophys Acta - Mol Basis Dis 2013;1832(1):46—56. Available from: https://doi.org/10.1016/j.bbadis.2012.10.010.

[28] Nie F, Yu X-L, Wang X-G, Tang Y-F, Wang L-L, Ma L. Down-regulation of CacyBP is associated with poor prognosis and the effects on COX-2 expression in breast cancer. Int J Oncol 2010;37(5):1261—9. Available from: https://doi.org/10.3892/ijo_00000777.

[29] Berger G, Leclercqz H, Derenne A, Gelbcke M, Goormaghtigh E, Nève J, et al. Synthesis and in vitro characterization of platinum (II) anticancer coordinates using FTIR spectroscopy and NCI COMPARE: a fast method for new compound discovery. Bioorganic Med Chem 2014;22(13):3527—36. Available from: https://doi.org/10.1016/j.bmc.2014.04.017.

[30] Das AK. Inorganic chemistry: biological and environmental aspects. Kolkata: Books and Allied (P) Ltd; 2004.

[31] Galloway JM, Bramble JP, Staniland SS. Biomimetic synthesis of materials for technology. Chem Eur J 2013;19(27):8710—25. Available from: https://doi.org/10.1002/chem.201300721.

[32] Bertini I, Gray HB, Lippard SJ, Valentine JS. Bioinorganic chemistry. Mill Valley, California: University Science Books; 1994.

[33] Kumar A, Kumar V. Biotemplated inorganic nanostructures: supramolecular directed nanosystems of semiconductor(s)/metal(s) mediated by nucleic acids and their properties. Chem Rev 2014;114(14):7044—78. Available from: https://doi.org/10.1021/cr4007285.

[34] Qu H, Caruntu D, Liu H, O'Connor CJ. Water-dispersible

iron oxide magnetic nanoparticles with versatile surface functionalities. Langmuir 2011;27 (6):2271–8. Available from: https://doi.org/10.1021/la104471r.

[35] Xiao L, Li J, Brougham DF, Fox EK, Feliu N, Bushmelev A, et al. Water-soluble superparamagnetic magnetite nanoparticles with biocompatible coating for enhanced magnetic resonance imaging. ACS Nano 2011;5(8):6315–24. Available from: https://doi.org/ 10.1021/nn201348s.

[36] Kumar A, Gupta SK. Synthesis of adenine mediated superparamagnetic colloidal β-FeOOH nanostructure(s): study of their morphological changes and magnetic behavior. J Nanopart Res 2013;15(1466):1–16. Available from: https://doi.org/10.1007/ s11051-013-1466-z.

[37] Kumar A, Gupta SK. Synthesis of 5′-GMP-mediated porous hydrogel containing β-FeOOH nanostructures: optimization of its morphology, optical and magnetic properties. J Mater Chem B 2013;1:5818–30. Available from: https://doi.org/ 10.1039/C3TB20877J.

[38] Wu H, Liu Z, Dong S, Zhao Y, Huang H, Zeng M. Formation of ferric oxyhydroxide nanoparticles mediated by peptides in anchovy (*Engraulis japonicus*) muscle protein hydrolysate. J Agric Food Chem 2013;61(1):219–24. Available from: https://doi.org/10.1021/jf3039692.

[39] Ouameur AA, Arakawa H, Ahmad R, Naoui M, Tajmir-Riahi HA. A comparative study of Fe(II) and Fe(III) interactions with DNA duplex: major and minor grooves bindings. DNA Cell Biol 2005;24(6):394–401. Available from: https://doi.org/ 10.1089/dna.2005.24.394.

[40] Qu X-F, Zhou G-T, Yao Q-Z, Fu S–Q. Aspartic-acid-assisted hydrothermal growth and properties of magnetite octahedrons. J Phys Chem C 2010;114(1):284–9. Available from: https://doi.org/ 10.1021/jp909175s.

[41] Gupta AK, Gupta M. Synthesis and surface engineering of iron oxide nanoparticles for biomedical applications. Biomaterials 2005;26

(18):3995–4021. Available from: https://doi.org/10.1016/j.biom aterials.2004.10.012.

[42] Deliyanni EA, Peleka EN, Matis KA. Effect of cationic surfactant on the adsorption of arsenites onto akaganeite nanocrystals. Sep Sci Technol 2007;42:993–1012. Available from: https://doi.org/ 10.1080/01496390701206306.

[43] Shin S, Yoon H, Jang J. Polymer-encapsulated iron oxide nanoparticles as highly efficient fenton catalysts. Catal Commun 2008;10 (2):178–82. Available from: https://doi.org/10.1016/j.catcom. 2008.08.027.

[44] Nakamura T. Acicular Magnetic Iron Oxide Particles and Magnetic Recording Media Using Such Particles. U.S. Patent 5120604, 1992.

[45] Gellert M, Lipsett MN, Davies DR. Helix formation by guanylic acid. Proc Natl Acad Sci USA 1962;48 (12):2013–18. Available from: https://www.ncbi.nlm.nih.gov/pmc /articles/PMC221115.

[46] Detellier C, Laszlo P. Role of alkali metal and ammonium cations in the self-assembly of the 5′-guanosine monophosphate cations. J Am Chem Soc 1980;102(37):1135–41. Available from: https://doi.org/ 10.1021/ja00523a033.

[47] Jurga-Nowak H, Banachowicz E, Dobek A, Patkowski A. Supramolecular guanosine 5′-monophosphate structures in solution. Light scattering study. J Phys Chem B 2004;108(8):2744–50. Available from: https://doi.org/ 10.1021/jp030905.

[48] Panda M, Walmsley JA. Circular dichroism study of supramolecular assemblies of guanosine 5′-monophosphate. J Phys Chem B 2011;115 (19):6377–83. Available from: https://doi.org/10.1021/jp201630g.

[49] Georgakilas V, Perman JA, Tucek J, Zboril R. Broad family of carbon nanoallotropes: classification, chemistry, and applications of fullerenes, carbon dots, nanotubes, graphene, nanodiamonds, and combined superstructures. Chem Rev 2015;115(4744):4744–822. Available from: https://doi.org/ 10.1021/cr500304f.

[50] Edwards RS, Coleman KS. Graphene synthesis: relationship to applications. Nanoscale 2013;5 (1):38–51. Available from: https://doi.org/10.1039/C2NR32 629A.

[51] Bonaccorso F, Colombo L, Yu G, Stoller M, Tozzini V, Ferrari AC, et al. Pellegrini V. Graphene, related two-dimensional crystals, and hybrid systems for energy conversion and storage. Science 2015;347(6217):1246501–9. Available from: https://doi.org/ 10.1126/science.1246501.

[52] Reina G, González-Domínguez JM, Criado A, Vázquez E, Bianco A, Prato M. Promises, facts and challenges for graphene in biomedical applications. Chem Soc Rev 2017;46(15):4400–16. Available from: https://doi.org/10.1039/ C7CS00363C.

[53] Cheng C, Li S, Thomas A, Kotov NA, Haag R. Functional graphene nanomaterials based architectures: biointeractions, fabrications, and emerging biological applications. Chem Rev 2017;117(3):1826–914. Available from: https://doi.org/ 10.1021/acs.chemrev.6b00520.

[54] Goenka S, Sant V, Sant S. Graphene-based nanomaterials for drug delivery and tissue engineering. J Control Release 2014;173:75–88. Available from: https://doi.org/10.1016/j. jconrel.2013.10.017.

[55] Tonelli FMP, Goulart VAM, Gomes KN, Ladeira MS, Santos AK, Lorençon E, et al. Graphene-based nanomaterials: biological and medical applications and toxicity. Nanomedicine 2015;10 (15):2423–50. Available from: https://doi.org/10.2217/nnm.15.65.

[56] Kumar A, Khandelwal M. A novel synthesis of ultra thin graphene sheets for energy storage applications using malonic acid as a reducing agent. J Mater Chem A 2014;2(47):20345–57. Available from: https://doi.org/10.1039/ C4TA04986A.

[57] Dimiev AM, Eigler S, editors. Graphene oxide: fundamentals and applications. 1st ed. West Sussex, United Kingdom: John Wiley & Sons Ltd; 2016. Available from: https://

www.wiley.com/en-us/Graphene + Oxide%3A + Fundamentals + and + Applications-p-9781119069409.

[58] Nanda SS, Papaefthymiou GC, Yi DK. Functionalization of graphene oxide and its biomedical applications. Crit Rev Solid State Mater Sci 2015;40(5):291−315. Available from: https://doi.org/10.1080/10408436.2014.1002604.

[59] Singh DP, Herrera CE, Singh B, Singh S, Singh RK, Kumar R. Graphene oxide: an efficient material and recent approach for biotechnological and biomedical applications. Mater Sci Eng C 2018;86:173−97. Available from: https://doi.org/10.1016/j.msec.2018.01.004.

[60] Liu Z, Liu J, Wang T, Li Q, Francis PS, Barrow CJ, et al. Switching off the interactions between graphene oxide and doxorubicin using vitamin C: combining simplicity and efficiency in drug delivery. J Mater Chem B 2018;6(8):1251−9. Available from: https://doi.org/10.1039/C7TB03063K.

[61] Kolanthai E, Sindu PA, Khajuria DK, Veerla SC, Kuppuswamy D, Catalani LH, et al. Graphene oxide—A tool for the preparation of chemically crosslinking free alginate-chitosan-collagen scaffolds for bone tissue engineering. ACS Appl Mater Interfaces 2018;10(15):12441−52. Available from: https://doi.org/10.1021/acsami.8b00699.

[62] Khandelwal M, Kumar A. One-step chemically controlled wet synthesis of graphene nanoribbons from graphene oxide for high performance supercapacitor applications. J Mater Chem A 2015;3(45):22975−88. Available from: https://doi.org/10.1039/C5TA07603J.

[63] Shareena TPD, McShan D, Dasmahapatra AK, Tchounwou PB. A review on graphene-based nanomaterials in biomedical applications and risks in environment and health. Nano-Micro Lett 2018;53(10):1−34. Available from: https://doi.org/10.1007/s40820-018-0206-4.

[64] Zhao H, Ding R, Zhao X, Li Y, Qu L, Pei H, et al. Graphene-based nanomaterials for drug and/or gene delivery, bioimaging, and tissue engineering. Drug Discov Today 2017;22(9):1302−17. Available from: https://doi.org/10.1016/j.drudis.2017.04.002.

[65] Kumar S, Raj S, Kolanthai E, Sood AK, Sampath S, Chatterjee K. Chemical functionalization of graphene to augment stem cell osteogenesis and inhibit biofilm formation on polymer composites for orthopedic applications. ACS Appl Mater Interfaces 2015;7(5):3237−52. Available from: https://doi.org/10.1021/am5079732.

[66] Zheng XT, Ananthanarayanan A, Luo KQ, Chen P. Glowing graphene quantum dots and carbon dots: properties, syntheses, and biological applications Small 2015;11(14):1620−1436. Available from: https://doi.org/10.1002/smll.201402648.

[67] Li K, Liu W, Ni Y, Li D, Lin D, Su Z, et al. Technical synthesis and biomedical applications of graphene quantum dots. J Mater Chem B 2017;5(25):4811−26. Available from: https://doi.org/10.1039/C7TB01073G.

# Chapter | 7 |

# Microscopy

*Ankur Baliyan[1], Hideto Imai[1] and Vinit Kumar[2]*

[1]NISSAN Analysis and Research Center, Yokosuka, Japan, [2]Amity Institute of Molecular Medicine and Stem Cell Research, Amity University, Noida, India

## 7.1 Introduction

Nanotechnology is an exciting new area in science and is considered to be one of the hottest topics in the 21st century. Everything in nature is made of microscopic or macroscopic objects, and we as humans are curious to understand the interesting properties of these objects, and at times, even wish to engineer the specific property that is best suited for a particular application. Predominantly, an object's properties depend on its structure and, in turn, its atomic arrangement. With the advancement in analytical techniques researchers now have power to explore more complex and intriguing structures that was considered unachievable few decades ago. The atomic world at the nanoscale is fascinating and to explore it requires some of the sophisticated analytical techniques such as transmission electron microscopy (TEM), scanning electron microscopy (SEM), atomic force microscopy (AFM), and confocal microscopy. TEM is an electron microscopy technique that magnifies the objects up to 10,000,000 times, to provide access to the internal structure: the atomic arrangement at nanoscale. Scanning electron microscope magnifies 500,000 times, and is used to analyze the surface/subsurface morphologies of any sample. AFM is a type of scanning probe microscopy (SPM), which has resolution to fractions of a nanometer, and provides surface information by touching the surface with a mechanical probe. Confocal is an optical imaging technique that provides very high spatial resolution and contrast compared with the conventional wide-field optical microscopy. Each technique has its own pros and cons, and at times they complement each other; often a single technique alone can't provide all the information needed

to understand the property of the desired sample. In this chapter, we discuss some of the experimental issues that a researcher must understand prior to starting with TEM, SEM, AFM, and confocal, and in addition, advanced data analysis techniques are also discussed.

## 7.2 Transmission electron microscopy

Transmission electron microscopy (TEM) is an electron microscopy technique that uses a beam of fast-moving electrons to illuminate a sufficiently thin sample (1–100 nm), to magnify the features (shape, size, and volumetric information) of the desired specimen. TEM is very similar to the optical microscope except that the former utilizes the electrons' source to illuminate the sample and the latter uses the light source. In an optical system, the radius of curvature of an optical lens (front and back) decides the focal length and magnification, however, in practice, there is a limit to increasing the magnification by using a single lens, and that is why the combination of two or more lenses comes to the rescue to achieve the desired magnification. Although, a combination of two or more lens provides the maneuvering ability and helps to fold the beam to make a compact optical setup, nevertheless, the aberrations are also compounded with the addition of more lens (aberrations). For an optical system, at higher magnification (100 times), the diffraction starts to dominate ($D = 1/\lambda$), limiting the resolving power of the optical system. Resolving power, the ability of an instrument to distinguish the two objects placed at a certain distance, is limited by the nature of the illumination source, that is, the

Data Processing Handbook for Complex Biological Data Sources. DOI: https://doi.org/10.1016/B978-0-12-816548-5.00007-1

wavelength of the light source, typically around 200 nm. In other words, by using the optical light source, one cannot distinguish the two objects placed adjacent to each other given the lateral distance between them is less than 200 nm. Unlike the light source, the wavelength of the electrons is much smaller; that is, $\lambda = 0.00251$ nm (microscope is operated at 200 keV), therefore higher resolution, depth of focus, and magnification can be attained [1].

From the instrumentation perspective, irrespective of any characterization technique, three things are common: (1) illumination source, (2) sample, and (3) detector. Based on the electromagnetic energy spectrum (EM) that ranges from radio to gamma wave (low energy to high energy), there can be varieties of illumination source and if a particular illumination source is selected then the detector must be selected accordingly; that is, if the illumination source is electron then the detector should be responsive to the electron. The fundamental change that happened because of the selection of the illumination source essentially decides the type of interaction (principle behind the characterization) between the illumination source and the specimen. In TEM, the basic principle is based on contrast or diffraction, when the electron beam passes through the sample a portion of the beam is

absorbed and a portion is transmitted through it forming an image on the phosphorescent screen or charge coupled device (CCD) [1,2].

Fig. 7.1 shows the schematic of TEM. Electrons from the electron gun are focused to the specimen using the series of an electromagnetic condenser lens, the electrons are absorbed or passed through the sample and form an image via projection lens onto the phosphorescent screen or CCD camera. One of the prerequisites for TEM is sufficiently thin (e.g., $1-100$ nm thickness) samples. In TEM, the illumination system employs either the thermionic or field emission electron gun to generate the beam of electrons. Thermionic emission (hot emission) gun uses tungsten electrode as a filament, which often takes $20-25$ minutes to stabilize. On the other hand, the field emission gun (cold emission) uses $LaB_6$ filament, which provides higher brightness, better coherence, and can quickly stabilize within $2-3$ minutes. A very good vacuum must be maintained in the TEM column to avoid the unwanted scattering of the incident electrons with the residual gas in the column; the presence of residual gas (impurities) can cause the sudden electron discharge, and at times, electron discharge can increase the electron beam current to such an extent that can trigger the TEM emergency shutdown [1].

**Figure 7.1** Illustration of optical and transmission electron microscope.

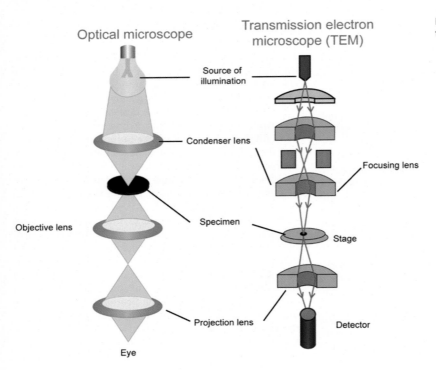

(A) 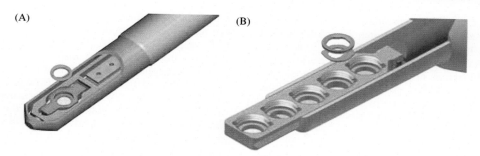 (B)

**Figure 7.2** (A) Illustration of single-sample holder for TEM. (B) Illustration for multiple-sample holder for TEM.

In general, TEM is operated at the voltage of the 80−300 kV; the selection of the operating voltage depends on the nature of the sample: soft or hard samples. Although the wavelength of the electron is inversely proportional to the energy of the electron, and in turn, gives higher resolution at higher energy, nevertheless, higher resolution is not the lone criterion, especially for biological samples where the size of the biological organelles is of a few μm. Unlike the nanomaterials, the biological samples are made of soft matter and can't withstand the higher energy electron beam of 200 kV or more, which is the reason the operating voltage for the biological samples and soft polymers should be around 80−100 kV. The interaction of the electron beam with the sample determines the contrast in the image. Interaction area is simply governed by the thickness of the sample and atomic number density of an element present in the sample, that is, elements with increasing atomic number have large atomic cross-section, and in turn, the higher absorption probability for electrons to be absorbed, which provides more contrast for sample imaging.

## 7.2.1 Sample preparation

For TEM, 2D and 3D sample holders are generally used as shown in Fig. 7.2. One can use a single-sample holder or multiple-sample holder. The single-sample holder can accommodate a single sample whereas a multiple-sample holder can accommodate many samples at a given time. Unlike the single-sample holder, multiple-sample holder is a hassle-free and high-throughput technique that minimizes the insertion and takeout attempt as many samples can be placed, however, the resolution, at times, is poor because of the fact that samples start to drift. For higher resolution imaging, we recommend using the single-sample holder to minimize the drift. The next section discusses in detail the sample preparation technique and issues associated with biological and nanomaterial specimens.

**Figure 7.3** Optical image of three blocks of molds, made from a special rubber, to produce flat blocks almost completely trimmed, ready for microtomy. Specimen material during embedding is easy and when hardened, these prenumbered blocks are easily removed from the mold by flexing it slightly. *Courtesy: Nissan ARC, Japan.*

### 7.2.1.1 Sample preparation: biological specimen

In TEM, one of the prerequisites to acquire quality images is an electron-transparent thin sample and that is why the specimen preparation becomes the crucial aspect of imaging. The biological cells, which are of the size of 10−40 μm, must be sectioned to make the sample thickness a few tens of nm and subsequently stained with heavy metals ($U_2O$) to give mass contrast. A typical process involved fixation, dehydration, and resin embedding of cells prior to the cell imaging via TEM; detailed protocol can be found elsewhere [3]. Fixation is a process where the cell is chemically treated and cell growth is instantly frozen, with glutaraldehyde, formaldehyde, and osmium tetroxide. In dehydration, the trapped moisture is removed because of the fact that the trapped moisture can cause the electrons to discharge and cause a nuisance during the TEM measurements. Finally, the dehydrated cells are embedded in resin to form a mold as shown in Fig. 7.3. Alternatively, TEM specimens can be prepared by ultrarapid freezing, in which the specimen is immersed into liquid ethane to drop its temperature at a rate of about $10^4$ °C s$^{-1}$ [4]. Quenching at such a higher rate is likely to freeze the living cells instantly because the sample is frozen within milliseconds. However, it is hard to visualize details of a sample at the cellular level because of the small differences in mass density between clusters of organelles; in addition, the information in frozen-hydrated samples is obscured by noise that results from camera performance and the poor electron counting statistics from the detector.

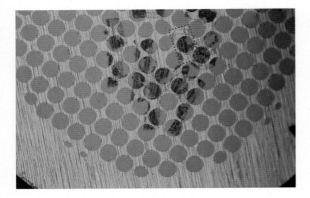

**Figure 7.4** Optical image of multiple thin slices of ultramicrotome sample placed on special TEM grid; the gray part indicates open space for observation and brown is opaque copper grid. One of the thin samples is marked with dashed red rectangle for better understanding.

For sectioning, there are two possibilities that one might opt for: either the ultramicrotome or focused ion beam (FIB), depending on the objectives. In ultramicrotome, the resin mold is sectioned into a very thin specimen using a wedge-shaped glass knife and the specimen is collected on the TEM grid for further investigation as shown in Fig. 7.4. However, 3D imaging artifacts may arise from the use of a blade to slice the sample, so the focused ion beam can overcome such kind of issue. FIB is a complex process in which a beam of gallium (Ga$^+$) is used to cut around 100 nm of the thin slice of constant thickness from the resin mold and subsequently transferred on to the TEM grid as shown in Fig. 7.5. The nanometer-scale resolution of the FIB allows the thin region to be chosen. However, the contrast of the cells will be very poor at this stage because of the fact that the cell is composed of very light atomic elements, mainly carbon, nitrogen, hydrogen etc., therefore, for better image contrast, it is advised to stain the cell specimen with heavy metal ions such as tungsten, uranyl, molybdate, or vanadate compounds prior to imaging [5].

Further, the selection of the TEM grid might also influence the image contrast and lead to a low-quality image. Although there are varieties of TEM grids available, which often causes confusion as to which one to select for a particular application. However, the rule of thumb is that the copper grid with or without any carbon film with a large pitch size of more than 5.0 μm can be the choice for the ultramicrotome samples [6], whereas for FIB sliced samples [7], the grids are very special and details can be

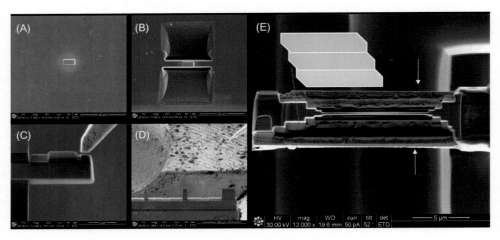

**Figure 7.5** Sequence of FIB process. Prior to observation, samples go through the following steps. (A) Desired sample to sectioned placed on substrate, (B) trenches are excavated near the sample, (C) trimmed sample attached to the TEM grid, (D) low resolution image of TEM grid after sample is attached to the grid, and (E) trenches are excavated from the sample using FIB gun (side and from top). Once the sample preparation is over sample can be investigated via TEM or SEM.
*Courtesy: Nissan ARC, Japan.*

found in the related literature [8]. It is recommended to wear antistatic gloves while preparing the sample to avoid any kind of static negative charge that might be transferred to the TEM grid, although, for lower magnification, it is not a severe problem. However, at higher magnification during the observation, the negative static charge on the grid and negatively charged electrons from the electron gun start to repel each other and the TEM grid starts to drift horizontally (causing the small lateral movement).

### 7.2.1.2 Sample preparation: nanomaterial samples

Unlike the biological samples, the sample preparation for nanomaterial (NMs) is a straightforward process. Since the NMs, zero (0D), one (1D), and two (2D) are ranged from 1 nm to few hundred nm, the TEM grid should have carbon film supported with microgrid framework. It is highly recommended to place either a tissue paper or filter paper below the TEM grid. A small drop of NMs dispersed in water or solvent is dropped on the TEM grid. The filter paper helps to absorb the excess amount of solvent. Given the carbon film on the TEM grid is hydrophobic in nature, at times, surface treatment of the TEM grid is desirable prior to sample casting on the grid. No treatment is required if either the intrinsic NMs or the functional groups on the surface of NMs are hydrophobic [9] as shown in Fig. 7.6. On the other hand, water-based hydrophilic NMs do not wet the surface of the TEM grid and often form a meniscus as soon as the sample is cast on to the TEM grid. For water-based NMs, it takes more time to dry and form large aggregates of the patchy island across the grid as shown in Fig. 7.7A. To avoid such issue, the TEM grid is subjected to the argon (Ar) plasma for 20−30 seconds prior to depositing the sample on the grid. Ar plasma etching removes some of the carbon from the surface of the TEM grid and transforms the surface from hydrophobic to hydrophilic, which assists in achieving a uniform distribution of NMs onto the TEM grid as shown in Fig. 7.7B.

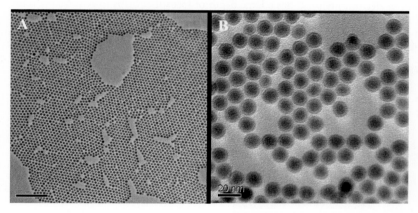

**Figure 7.6** TEM images of iron oxide core−shell nanoparticles: (A) at low resolution, (B) at higher resolution. *Courtesy: Bio-Nano Electronics Research Center(BNRC), Toyo University, Japan.*

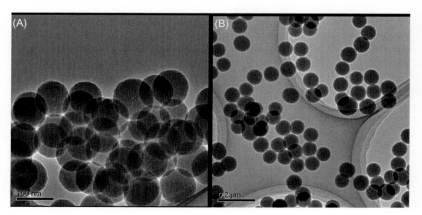

**Figure 7.7** (A) TEM image of silicon oxide ($SiO_2$) porous nanoparticles (the patch of nanoparticles can be seen), (B) TEM image of long-range disperse silicon oxide ($SiO_2$) porous nanoparticles. *Courtesy: BNRC, Toyo University, Japan.*

### 7.2.2 Three-dimensional tomography

Three-dimensional tomography is a technique where the TEM grid holder has an extra input circuitry to rotate the holder a certain amount by applying the external voltage as shown in Fig. 7.8 [10]. Often, to investigate the drug delivery using NMs, it is extremely difficult to differentiate from the 2-D TEM images whether the nanoparticles are present in the resin or lying on the surface of cell sections. Three-dimensional (3-D electron tomography) can help to ascertain the object shape, morphology, and whether the NM is internalized by the cell or not. Acquisition of images with the specimen tilted at multiple angles over a wide angular range (−70 to +70 degrees), followed by special tomographic reconstruction technique, render the reconstructed volumetric data information section by section [4].

## 7.3 Scanning electron microscopy

In SEM, unlike TEM, a low energy beam of electrons, typically between 1 and 30 keV, is focused into a narrow probe and scanned in a raster pattern across the surface of a sample. The electron beam interacts with the desired sample much similar to a teardrop shape, generally referred to as the interaction volume, as depicted in Fig. 7.9. Depending on the depth of interaction between the primary electrons and specimen surface (subsurface), different types of signals are generated, which extend from a few nm to 5.0 μm. The energy exchange between the electron beam and the specimen results in the reflection of high-energy backscattered electrons by elastic scattering (backscattered imaging mode), emission of secondary electrons by inelastic scattering (secondary electron imaging mode), and the emission of electromagnetic radiation (continuous and characteristic EDX Mode), each of which can be separately detected by respective specialized detectors [11].

Fig. 7.10 shows the schematic of SEM. The condenser lenses help to demagnify the electron source into a small probe and focus on the specimen surface via the objective lens. Using the scan coils (x- and y-axes) the probe is rastered from the top down onto the specimen and the final image is formed. Unlike the optical microscope, SEM has a large depth of focus that helps to observe the rough sample surface. The resolution of SEM depends on the type of detection mode (BSE or SE mode). In BSE mode, the probe size depends on the operating beam voltage

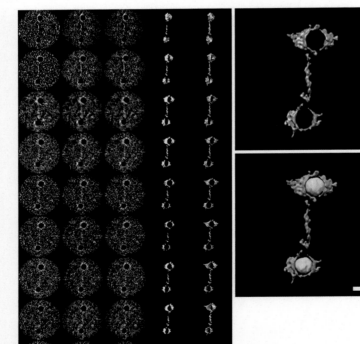

**Figure 7.8** 3D reconstruction of DNA—nanogold conjugate. Left panel: nine representative tilt images of the first targeted individual particle are displayed in the first column from the left (SNR of DNA portion: B0.31). Using IPET, the tilt images (after CTF correction) were gradually aligned to a common center for 3D reconstruction via an iterative refinement process. The projections of the intermediate and final 3D reconstructions at the corresponding tilt angles are displayed in the next four columns according to their corresponding tilt angles. Right upper panel: final IPET 3D density map of the targeted individual particle (SNR of DNA portion: B2.44). Right lower panel: the final 3D density map and its overlaid 3D density maps (final map in blue and its reversed map in gold) indicated the overall conformation of the DNA—nanogold conjugates. [10].

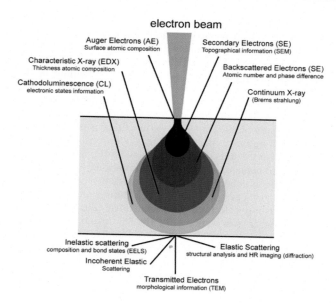

**electron beam**

Auger Electrons (AE)
Surface atomic composition

Secondary Electrons (SE)
Topographical information (SEM)

Characteristic X-ray (EDX)
Thickness atomic composition

Backscattered Electrons (SE)
Atomic number and phase difference

Cathodoluminescence (CL)
electronic states information

Continuum X-ray
(Brems strahlung)

Inelastic scattering
composition and bond states (EELS)

Elastic Scattering
structural analysis and HR imaging (diffraction)

Incoherent Elastic
Scattering

Transmitted Electrons
morphological information (TEM)

**Figure 7.9** Illustration of SEM electron beam profile after substrate interaction. Various interactions are highlighted.

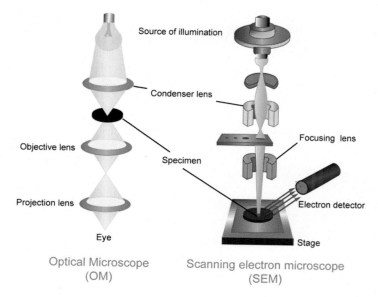

Source of illumination

Condenser lens

Objective lens

Focusing lens

Specimen

Projection lens

Electron detector

Eye

Stage

Optical Microscope
(OM)

Scanning electron microscope
(SEM)

**Figure 7.10** Illustration of optical and scanning electron microscope.

and optics, for high-resolution images the beam voltage of more than 10 kV is used. However, with an increase in the operating voltage, the interaction volume is also changed. In the secondary electron detection mode, an increase in operating voltage can significantly shift the secondary electron interactive area (interactive area is of tens of nm from the surface), towards the subsurface resulting in the poor spatial resolution, whereas for the backscattered detection mode the spatial resolution will improve given the large interactive area which is of the order of micrometer from the surface [12]. Although varieties of SEM are available [13], we will restrict the discussion to the conventional SEM, high-resolution SEM and environmental SEM. The conventional and high-resolution SEM are commonly used to observe the inorganic and hard nanomaterial, whereas the environmental SEM is used for the soft biomaterials and polymers.

The illumination system for SEM, similar to the TEM, employs either the thermionic or field emission electron gun to generate the beam of electrons. In conventional SEM, electron gun, detector, and other electro-optics are enclosed into a vacuum chamber to avoid any kind of unnecessary interaction with the residual gases that can cause the electron discharge. Usually, the low-cost SEM has a single vacuum chamber, whereas the high-end SEM has two vacuum chambers: an outer and inner chamber. One of the disadvantages with the one chamber instrument is that it consumes more time to evacuate the air from the chamber and repeatedly opening and closing of the SEM chamber unit might allow contamination over a period of time. In contrast, the inner chamber of the two chamber SEM is constantly under vacuum, which helps to avoid any kind of contamination and the small size outer chamber quickly evacuates. It is used for exchanging the sample. Furthermore, the outer chamber can act as a buffer to protect the sample to avoid environmental exposure, that is, under the special protected environment (N2 and Ar). Two important detection modes in SEM, backscattered electron mode (BSE) and secondary electron mode (SE), are shown in Fig. 7.11.

In backscattered electron mode, the primary incoming electrons from the electron beam are elastically scattered by atoms from the sample, which results in the change of trajectory of primary electrons and these reflected ones are designated as the backscattered electrons. The most common backscattered electrons reflect back at a low angle, to very close proximity to the electron gun. Depending on the beam energy and sample material, the backscattered electrons primarily generate from subsurface interaction volumes ($>100$'s nm). The BSE detector, a solid state detector that typically contains $p-n$ junctions, is placed near the incoming beam; the backscattered electron is absorbed by the detector and creates an electron—hole pair. The number of backscattered electrons are proportional to the average atomic number (scattering cross-section) of the sample, large scattering cross-section of heavy atomic number reflects more electrons as compared with the light elements of smaller cross-section [13]. The BSE images contain compositional contrast that can be utilized to distinguish distinct phases present in the sample. Fig. 7.12A shows the SEM image of biological cell observed in BSE mode.

In secondary electron imaging, the inelastic scattering of the primary electrons with the sample surface leads to ejection of low energy ($<50$ eV) secondary electrons from within a few nanometers of the specimen surface, that is, near the surface. Unlike the backscattered electron, the secondary electron reflects at a larger angle with respect to the incoming electron beam making it easy for the detection. The Everhart—Thornley detector is regularly used for the detection of secondary electrons. SE detector counts

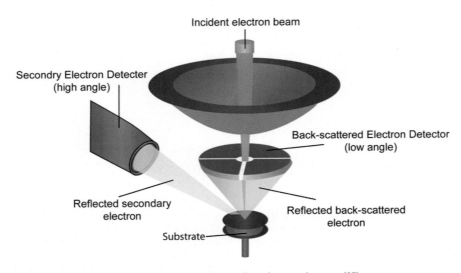

Figure 7.11 Illustration for backscattered detector (BSE) and secondary electron detector (SE).

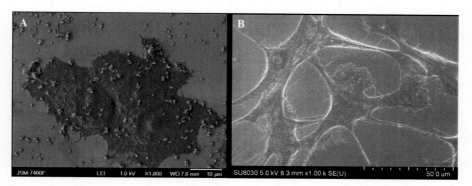

**Figure 7.12** SEM image of biological cell: (A) backscattered mode (BSE), (B) secondary electron mode (SE).
*Courtesy: BNRC, Toyo University, Japan.*

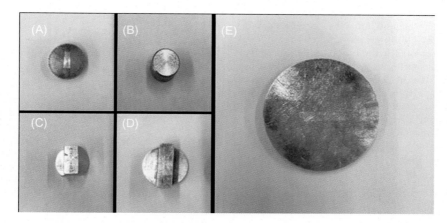

**Figure 7.13** Various kind of sample holder for SEM. (A and B) single-sample holder (1 cm diameter) made of aluminum and brass respectively, (C) One-side slant sample holder to observe the tilted cross-section, (D) two-side vertical sample holder for cross-section observation, (E) multiple-sample holder that can accommodate multiple sample (2 in. diameter).
*Courtesy: BNRC, Toyo University, Japan.*

the secondary electrons as the electron beam rasters across the sample using the photomultiplier tube [13]. However, for the secondary electron emission, the surface of the specimen must be sufficiently conductive, nevertheless, for biological and organic samples the conductivity is not sufficient enough to eject the secondary electrons, and thus, the biological samples must be coated with a thin layer of conducting substrate prior to the SEM observation (coating is discussed in more detail in the subsequent section) unlike the conductive samples such as metallic

CNTs, Fe NPs. Fig. 7.12B shows the SEM image of biological cell observed in SE mode. SEM with the large depth focus reveals the finer characteristics of the sample topography and surface morphology.

## 7.3.1 Sample preparation

There are varieties of sample holders available for SEM observation as shown in Fig. 7.13. One can use the broad sample holder capable of accommodating multiple samples or a smaller one to observe one by one. A vertical sample holder as shown in Fig. 7.13C is commonly used to observe the cross-section of the sample. A strong bonding is required between the sample and holder to avoid the sample spill from the holder while transferring the sample in and out from the SEM chamber. To avoid such kind of issues carbon paste or carbon tape is used to hold the sample firmly onto the holder. After the sample preparation is done, the sample can be moved into the SEM chamber for observation. However, depending on the sample characteristics such as softness or hardness, conducting or nonconducting, additional resources may be required.

### 7.3.1.1 Sample preparation: biological specimen

Biological samples cannot be observed directly using SEM because they are nonconducting and have lots of moisture trapped inside. The sample must be completely dehydrated prior to the SEM observation, free from any moisture or trapped solvent. The biological cells must be fixed with glutaraldehyde and postfixed with osmium tetroxide, followed by dehydration using ethanol, and dried at the critical point using carbon dioxide ($CO_2$) [14]. The diffusivity of the carbon dioxide at a critical point reaches to infinite and helps to penetrate into the cell for the complete removal of the moisture from the cell. Finally, to provide much-needed conductivity, the cells can be coated with Au, Pt, or Osmium.

#### 7.3.1.1.1 Coating the specimen surface

Polymer and biological samples are nonconducting and incoming electrons can accumulate on the surface of the sample, commonly known as the charge-up, and hinder the sample observation by creating multiple charged hotspots across the surface as shown in Fig. 7.14. Charge-up is more prominent at the sample edges, as primary electrons continuously strike on to the sample, and the excess electrons get accumulated and find no conducting path to get neutralized. The sample can be placed either on the carbon paste or carbon tape to avoid the sample being charged up. However, coating the semi- or nonconducting samples, including resin embedding ultramicrotome and FIB specimen, with a thin layer of conducting material, is a better alternative to avoid the charge-up. The specimen surface can be coated with gold, platinum, and carbon, which improves secondary electron yield and provides a conduction path to dissipate the charge and heat, minimizing the beam damage [15].

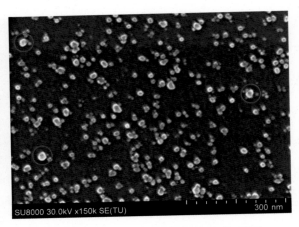

SU8000 30.0kV x150k SE(TU)　　　300 nm

**Figure 7.14** Charging of iron nanoparticles during SEM observation (edge effect), red circle is drawn around a charged nanoparticle for better understanding.
*Courtesy: BNRC, Toyo University, Japan.*

Generally, sputter coaters are employed to coat the soft material, including the biological samples, with gold, platinum, or osmium. The sample that is to be coated is placed on the coater holder and desired metal substrate (Au, Pt, Os) is bombarded with plasma, resulting in a thin layer deposition on to the sample surface. The thickness of the film is directly proportional to the time and power of the plasma source. Usually, 30−60 seconds of plasma is enough to make a coating between 5 and 50 nm; one must be careful enough not to coat excessively because the coated film might break down during SEM observation and form an inhomogeneous island of nanoparticles. Although selection of the coating material depends on many factors, nevertheless, it is recommended that gold or platinum can be chosen for conventional SEM. On the other hand, osmium coating is best suited to withstand the higher current density of primary electrons in FESEM. Osmium being a much heavier element than gold and platinum provides better contrast with a very thin and structureless coat.

A metal coating is also not suitable for those nanoparticles (NPs) that are functionalized with organic ligands. In addition, the breakdown of metal thin film coated on functionalized nanoparticles (due to the SEM electron exposure) creates a scenario where it is difficult to distinguish that the observed nanoparticles are from the sample or from the coating itself. Carbon thin film coating that possesses good electrical conductivity and low background signal might help the high-resolution imaging of functionalized nanoparticles (HRSEM) as can be seen in Fig. 7.15. An ancillary unit of carbon coater is often attached to the SEM; 30−60 seconds of plasma is enough to coat a very thin carbon film of around 5 nm onto the sample.

### 7.3.1.2 Sample preparation: nanomaterial specimen

Sample preparation for nanomaterials (NMs) is a straightforward process. The sample preparation strategy depends on the physical state of the NM. Solution based NMs can be uniformly coated on to a small silicon wafer either by spin coating or drop casting as shown in Fig. 7.16. For the NMs that are not in solution such as CNTs, a small

amount of NM picked using a toothpick can either be placed directly on carbon tape or dispersed in ethanol and then spin coated. At times, as-grown CNTs sample can be directly placed into the chamber for observation as shown in Fig. 7.17 [16].

### 7.3.2 Environmental scanning electron microscope

Unlike the conventional SEM, the environmental scanning electron microscopy (ESEM) does not require any additional sample preparation steps, and the sample can be observed in low-pressure gaseous environments (1−50 Torr). The development of a secondary-electron

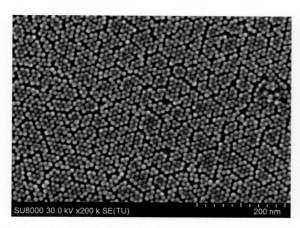

**Figure 7.15** SEM image of oleic acid functionalized iron oxide nanoparticle deposited on silicon substrate after carbon coating.
*Courtesy: BNRC, Toyo University, Japan.*

**Figure 7.17** SEM image of carbon nanotubes grown on silicon substrate, CNTs were grown on 2 in. silicon wafer (broad substrate SEM holder was used for investigation).
*Courtesy: BNRC, Toyo University, Japan.*

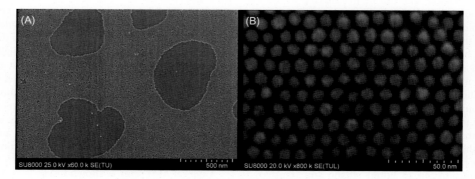

**Figure 7.16** (A) Low- and (B) high-resolution SEM image of long-range iron oxide nanoparticle deposited on silicon substrate via spin coating.
*Courtesy: BNRC, Toyo University, Japan.*

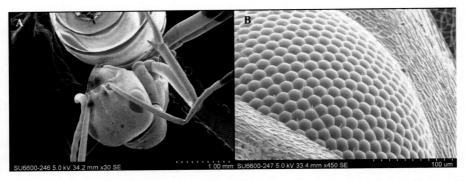

**Figure 7.18** Low- and high-resolution image of an ant taken by environmental SEM, (A) low resolution of an ant, (B) high resolution of ant eyeball.
*Courtesy: BNRC, Toyo University, Japan.*

detector has made it possible to operate the SEM in the presence of limited water vapor by the use of pressure-limiting apertures. ESEM is operated at a low accelerating voltage (e.g., 1–2 kV) that can limit the resolution as shown in Fig. 7.18. However, ESEM has the advantage over conventional SEM because the insulating samples do not require any coating prior to the observation. Occasionally, coating might obscure the fine structure of the specimen surface, and because of the high-energy plasma ions there is always a high risk of sample surface structure alteration.

### 7.3.3 Three-dimensional scanning electron microscope

Traditionally, TEM has been used for higher resolution structural studies. However, the practical limit for the thickness of a sample that can be investigated by TEM is less than 500 nm; thicker samples lead to a decrease in the information content. Thus large samples at high resolution with thicknesses on the order of micrometers are investigated by 3D SEM. Cell and tissue architecture, as preserved in embedded resin or in frozen form, can be investigated implying 3D SEM by progressive removal of material using a focused ion beam. Recently developed rotary microtome technique generates ribbons of sequential sections that are continuously collected on an adhesive strip, attached to a large wafer, and subsequently imaged by the scanning electron microscope as depicted in Fig. 7.19 [17]. Alternately, in another FIB technique, cells are plunge-frozen in liquid ethane and transferred to a FIB-SEM. A chosen area is FIB-milled tangentially either from the top or side revealing the fresh region, still encased in vitreous ice, imaged by FIB-SEM to

produce the 3D volumetric reconstruction of cell architecture, as shown in Fig. 7.20 [17].

## 7.4 Atomic force microscopy

Atomic force microscopy is a novel technique for high-resolution imaging of conducting and nonconducting surfaces. The interaction between the specimen surface and sharp AFM probe results in surface information of the specimen. There can be two types of interactions between the sample and the tip: short range and long range. In short range interactions, the van der Waals forces dominate whereas in long-range interactions, electric and magnetic forces dominate. Here, the long-range forces are not discussed and details can be found elsewhere [18]. The foundation of AFM is the potential energy versus distance curve as observed in Fig. 7.21. The net force between the two atoms is the result of attractive $(1/r^2)$ and repulsive $(1/r^6)$ forces, however, the slope of repulsive force is more dominating at shorter distances than the attractive force. Consider the two atoms, at far distance; the net force between the two atoms is almost negligible, as the distance between the two atoms starts to decrease the forces become attractive, as one moves closer to the other the attraction forces reaches maximum at the interatomic distance between the two atoms, after that repulsive forces start to dominate and continue to be more repulsive. In AFM, the interaction between the probe and specimen surface atoms is observed through deflection, which is associated with the amount of force applied.

A typical AFM setup is shown in Fig. 7.22. It has a stage that can move in $x$-, $y$-, and $z$-direction. The sample is mounted on the stage and probed using a cantilever, depending on the interaction between the cantilever and

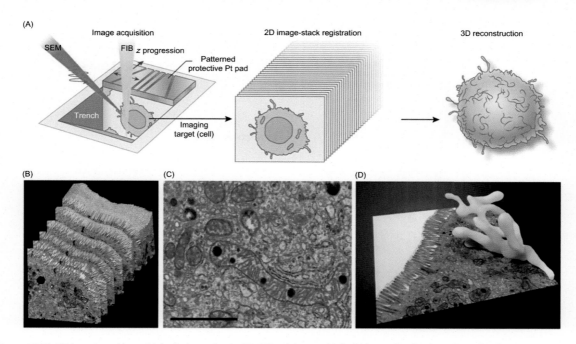

**Figure 7.19** 3D imaging of large biological samples by FIB-SEM. (A) Large biological samples that have been fixed either conventionally (by aldehydes) or cryogenically (by high-pressure freezing), stained by heavy metals, resin embedded, and mounted are introduced into the FIB-SEM chamber. Here, chosen areas of the sample are "trenched" to reveal the region of interest and then subjected to an iterative cycle of resin milling by the FIB (*yellow beam*) followed by SEM (*blue beam*) imaging of the newly revealed face to produce a 2D image stack. The patterned protective platinum (Pt) pad atop the sample to be imaged allows automatic beam tuning and slice-thickness control. The 2D image stack is then computationally converted to a 3D volume, aligned, and segmented to reveal the 3D structure of interest. (B—D) A representative example of 3D tissue imaging using a mouse intestinal sample. Shown are an image stack (B), a selected slice through the stack (C), and a segmented representation of an extensively branched mitochondrion present in the imaged volume (D). Scale bar, 1 mm. Panels (B—D) reprinted from *Encyclopedia of Cell Biology*, Vol. 2, Hartnell, L.M. et al., "Imaging cellular architecture with 3D SEM," 44—50, Copyright 2016, with permission from Elsevier [17].

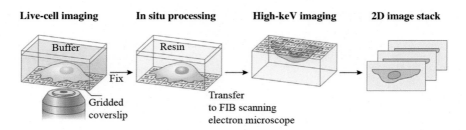

**Figure 7.20** 3D imaging of specific targets with correlative LM and FIB-SEM. LM of a biological sample grown on or attached to an alphanumerically coded gridded coverslip produces a "coordinate map" whose fidelity is maintained after resin embedding in situ, allowing location of the ROI for FIB-SEM imaging [17].

sample surface, the force—distance curve is measured in terms of deflection. A displacement sensor is utilized to measure the deflection caused by the surface interaction between the probe and specimen surface. The laser light is incident on the backside of the AFM tip, which is coated with a very good reflective surface. As the cantilever comes into contact with the specimen surface, the interaction at a given time might be attractive or repulsive as the probe

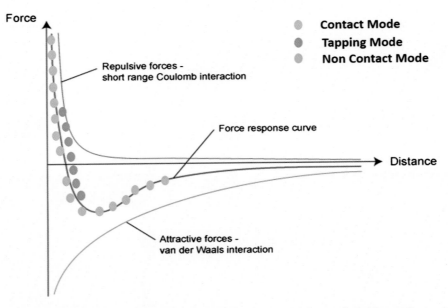

**Figure 7.21** Force–distance curve illustration; depending on the force curve various types of scanning mode are also highlighted.

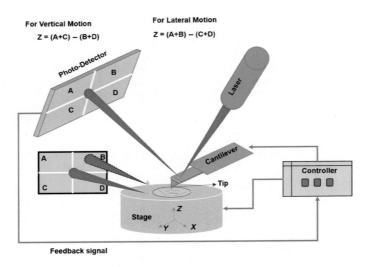

**Figure 7.22** Illustration of AFM set-up is shown. Laser beam is reflected from the cantilever and detected via photodetector, deflection is measured for lateral or vertical motion, and a feedback signal is sent to the controller to adjust the height accordingly.

move across. The deflection ($Z$) is measured from the central position of the photodetector. Any change in the surface morphology results in the shift (deflection) in the reflected laser beam from the central position and this deflection serves as a feedback to the processing unit to maintain the specified distance between the cantilever and sample, which in turn provides the specimen surface morphology.

For AFM, there are three modes available for imaging the sample surface: contact, noncontact, and tapping modes. In contact mode, the probe is always maintained in contact with the sample surface with the desired set point (voltage) and the surface is modeled from the deflection of the cantilever. In noncontact mode, the probe, vibrating with a certain frequency, is not in direct contact with the sample surface, rather uses a soft set point (voltage) and the change in the vibration amplitude is modeled and surface contours are plotted. In tapping mode, the cantilever tip is periodically in contact with the sample surface, and the surface is drawn based on the change in the vibration amplitude of the oscillating cantilever. Noncontact and tapping modes are widely used because of the fact that the tip is not constantly in touch with the surface, whereas in the contact mode the strong repulsive forces can damage the sample surface.

The special resolution of AFM is limited by the probe cantilever. Sharper cantilever tip results in the high-resolution image of nanoscale surface morphology. Nowadays, fast imaging AFM is available, which can essentially complete the scan within minutes with high-resolution precision [19]. Widely used cantilevers are Si and Si3N4. The optical sensors measure the deflection of the cantilever with a sensitivity of 0.1 nm. The selection of an appropriate cantilever can be quite tricky; there are varieties of cantilever probe readily available on the market [20]. There are multiple factors that affect the probe selection: sample surface, environment (liquid or air), etc. The probe selection largely depends on whether the sample is soft or hard. The cantilever probe with large force constant (hard tip) can damage the surface of soft material such as polymer.

## 7.4.1 Sample preparation

The imaging of sample surface using the AFM provides superior structural and local higher resolution information compared with the conventional electron microscopy (EM), X-ray crystallography (X-RD), and nuclear magnetic resonance (NMR). Unlike electron microscopy (EM), AFM does not require any vacuum column and does not impose any restriction that samples need to be conducting. However, a large chunk of time goes into the probe mounting on the AFM probe holder and subsequently the optical alignment: tip position and laser beam. The probing cantilever not only detects the chemical force but also long-range interactions such as van der Waals forces and electrostatic forces. AFM is also very sensitive to the noise; for example, doors closing, movement of humans, and even the flow of the air conditioner can leave their signature in the form of noise while scanning. To minimize the noise AFM should be enclosed in the box with a vibration isolation table. The sample preparation for biological and conventional NM is discussed in the following sections.

### 7.4.1.1 Sample preparation for biological

Probing the biological sample with atomic force microscopy is considered to be challenging. The force interaction is substantially complex because of the average size of animal cell ranging from 10.0 to 40.0 μm. Immobilization of biomolecules becomes critical to the kind of base substrate being used. The base substrate itself should not interact with the biological sample strongly to cause any unwanted interaction other than what is being probed and should not be loosely coupled such that the sample starts to drift as the probe scans across the sample. The biological sample should be firmly attached to the substrate (glass, coverslip, mica), a spherical animal cell is easier to detach from the glass surface than nonspherical ones. The larger the contact surface area between the sample and the substrate the better is the immobilization of biomolecules. In general, biological samples can be investigated after fixation onto the substrate. Fixation means that the cells are placed on the desired substrate and the sample is allowed to dry in a dust-free environment, subsequently investigated by AFM either in contact or noncontact mode. It is advised to avoid the contact mode specifically for those biological samples that are very soft.

AFM can be done in a liquid environment, which helps to investigate the dynamics of living cells in real time. Recently, AFM is becoming faster and faster as the software and hardware architecture of the computer improved significantly. For the cell dynamics observation, one should be able to scan the biological sample before the cell changes its structure, that is, imaging speed should be fast enough to investigate the cell almost in real time. In dynamic mode, the sample is fully immersed in the liquid environment and scanned in tapping mode by the force modulation. The hardness map can be obtained by mapping the cell in phase, modulus, and height mode. One of the prerequisites for the cell dynamic observation is that the pH of the culture media should be constantly maintained either by perfusing the cell culture or exposing the chamber to $5\%CO_2/95\%$ air [21].

A cell adheres to the sample in a medium inside a dish kept on the AFM stage. Slowly the probe is brought closer to the cell and the lateral load is applied to the cell. The cantilever is deflected corresponding to the surface of the cell morphology and deflection is plotted in the three-dimensional contour, in turn giving the (3D) surface morphology of the sample. The shape of the DNA, collagen, and DNA mobilized on the CNTs surface is presented in Fig. 7.23 [22]. DNA sometimes does not bind to

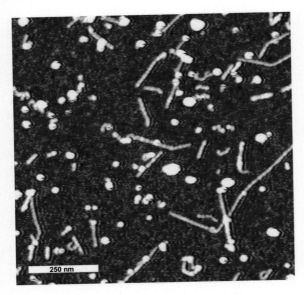

**Figure 7.23** AFM height image of CNTs wrapped with $T_{60}$ oligonucleotide, showing the surface pattern along the length of the CNTs and its prevalence (5-nm scale) [22].

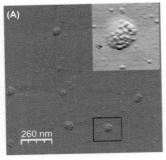

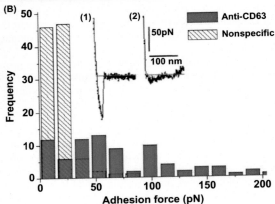

**Figure 7.24** Biochemical characterization of exosomes via AFM immunogold imaging and force spectroscopy showing the presence of CD63 receptors on the exosome surface. (A) Multiple CD63 receptor sites identified with anti-CD63 monoclonal antibodies and secondary antibody-gold beads. AFM topographic image showing 5–8 nm functionalized beads bound specifically to exosomes. The inset shows a zoomed out 3D image of individual beads bound to the surface of an isolated single exosome. (B) Distribution of rupture events. Nonspecific interactions occur mostly at forces <50 pN, while specific CD63 antibody-induced forces were distributed in the range of 30–200 pN. Sampled forces ($n = 80$ each) had each data point representing a single force measurement at any position on the exosomes surface (bin size 15 pN). Typical curves showing force (pN) as a function of separation (nm) for a single pull with (1) strong adhesive event between anti-CD63 and (2) no event for nonspecific antibody functionalized tip and exosome [23].

the mica, which is because freshly cleaved mica and DNA both have the same negative charge that causes the cells to loosely bind. Treating the mica with divalent and trivalent cations can provide much-needed immobilization force between the cells and mica.

The hydrodynamic effect must also be considered prior to the investigation; the free flow of fluid around the cantilever is disrupted by the presence of the nearby surface. In addition, as the probe touches the sample the cantilever may tilt slightly, in turn sensing the torsion force rather than the lateral force, so the accuracy of the data cannot be guaranteed. At times, the unnecessary fragments of biomolecules often get attached to the surface of the scanning tip during the scanning of the sample resulting in poor resolution or completely obscuring the surface morphology of the sample. To avoid such kind of scenario, it is recommended that the tip is lifted off and tuned to check the frequency of the tip; if the tip frequency does not fall into the recommended natural frequency of the tip mentioned by the manufacturer, then either the tip must be changed or it should be cleaned prior to observation. At times, the probe tip is functionalized with biomolecules such as DNA or antibody [23] and scanned through the specimen surface to estimate the force constant between the different types of biomolecules as shown in Fig. 7.24.

### 7.4.1.2 Sample preparation for nanomaterial

AFM is very sensitive to the change in surface morphology and widely used to scan a variety of nanomaterials. We have observed many kinds of nanomaterial starting from zero, one, and two dimensional NM. The sample preparation is almost similar to that mentioned in

Section 7.2.1.2. The most challenging part is the selection of the base substrate on which the sample can be deposited. Most of the base substrate has an intrinsic surface roughness (not flat) and if the base substrate itself is not flat enough, one might expect variation in the height of the NM, giving erroneous data. Iron oxide monodisperse nanoparticles (NPs) dispersed in hexane were spin coated and deposited on the silicon substrate as presented in Fig. 7.25. The bare silicon substrate itself has a surface roughness of 0.21 nm. Freshly prepared silicon substrate gets oxidized slowly either in the open environment or during the AFM observation, and surface roughness gradually increases from 0.21 to 31 nm after 24 hours. On the other hand, HOPG has a very smooth surface and can be a good alternative as the base substrate. Iron oxide nanosheets deposited on the silicon substrate using spin coating are shown in Fig. 7.26. They have three-dimensional (3D) interconnected architecture with lots of pores.

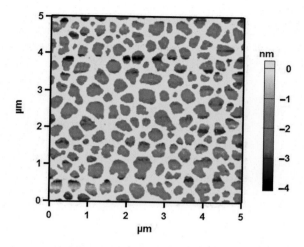

**Figure 7.25** AFM image of a single layer of iron nanoparticle deposited on the silicon substrate via spin coating. Average size of the nanoparticles was 3.88 nm, which also exactly matched the height profile of the AFM image.
*Courtesy: BNRC, Toyo University, Japan.*

## 7.5 Confocal microscopy

Confocal microscopy is an optical imaging technique that provides very high spatial resolution and contrast compared with the conventional wide-field optical microscopy with additional advantages such as control over field depth, minimal background signature, and ability to collect serial optical sections from thick specimens [24]. A conventional microscope is marred with multiple depth of field; an optical lens has a region (depth of field) at focal length f where the image can be formed and within this region, multiple images are formed. The superimposition of multiple images at the detector because of the multiple focal planes resulted into a blurred image. In general, a confocal microscope can be classified into noncoherent confocal microscope and laser confocal microscope. A noncoherent confocal microscope uses mercury or xenon as a light source, and optical color filters can be inserted to the light beam to choose specific wavelength for sample illumination. On the other hand, a single wavelength or multiwavelength laser is used in laser confocal microscope to illuminate the sample. The laser as a light source

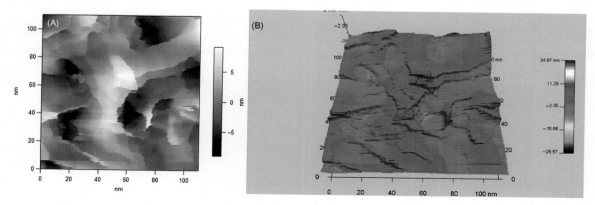

**Figure 7.26** (A) AFM image ultrathin Fe nanosheets, (B) 3D architecture of Fe nanosheets (height mode).
*Courtesy: BNRC, Toyo University, Japan.*

provides much better contrast and coherence needed for quality images.

Recently, the confocal microscope has become an indispensable tool for the biologist to study cell structures and their functions. Confocal provides unprecedented microscopic details of cells and tissues' complex morphology and dynamics with extremely high-quality superresolution images. Fig. 7.27 shows the precise expression of E cadherin on the cell junctions of pancreatic cancer cells. It is worthy to mention that though TEM offers excellent resolution up to the atomic level, but the beam of high-energy electrons can damage the living specimens; in addition, TEM sample preparation can potentially cause some artifacts from fixation and sectioning, while these limitations are addressed by confocal.

In conventional microscopy the sample is uniformly illuminated with high-intensity light beam and the objective lens, which is focused at the required focal plane of the sample, collects the reflected photons from the sample surface. This process produces out-of-focus blur images from areas above and below the focal plane. Whereas, in confocal microscopy, the unwanted out-of-focus light is eliminated by introducing the small aperture: a very small spot or a series of spots [25]. A spatial pinhole is used to cut most of the out-of-focus light during image formation, which results in sharply defined object plane devoid of out-of-focus blur. Baligar et al. have studied engrafting of bone marrow (BM)-derived cells in the liver and their proliferation [26]. Confocal microscopy analysis shows that the eGFP expressing cells assumed hepatic morphology and most of them express albumin (Fig. 7.28).

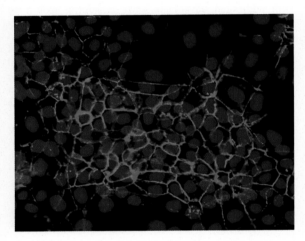

**Figure 7.27** Pancreatic ductal adenocarcinoma (PDAC) cells stained with E cadherin antibody. FITC is used as secondary antibody. Images were taken from Nikion confocal microscope. The expression of E cadherin on the cell junctions of pancreatic cancer cells is clearly visible.

## 7.6 Data analysis: transmission electron microscopy, scanning electron microscope, atomic force microscope, and confocal microscope

TEM, SEM, AFM, and confocal measurement results in plenty of images—a large data set. Data can be either a single image, sequence of images from the same sample, or the integrated image data set from the discrete samples. Data management and analytics become essential to extract the qualitative and quantitative information. Mathematically, an image can be translated into a matrix and lots of operations can be performed on the images like contrast and brightness enhancing, feature extraction, noise filtration, size or pore distribution via histogram, thickness estimation, volumetric and surface quantification, etc. Although most of the instruments have embedded image analytics software, nevertheless, the inbuilt software is capable of only basic operations and does lack

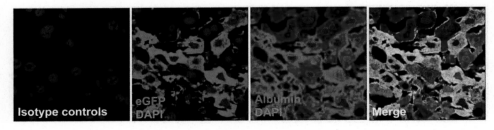

**Figure 7.28** Engrafting of bone marrow-derived cells in liver and their proliferation. Confocal micrographs of engrafted donor cells coexpressed eGFP and albumin. Magnification = ×630 [26].

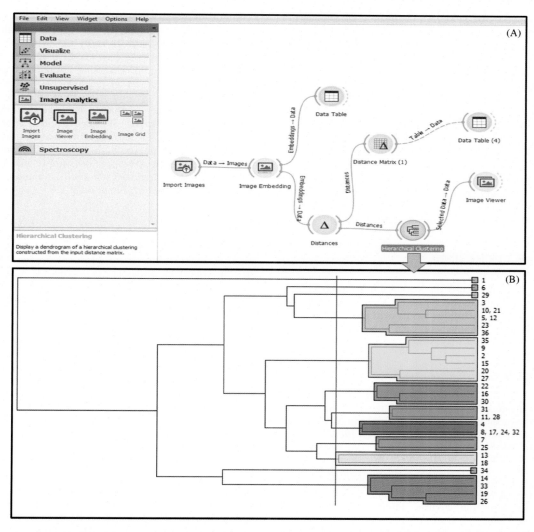

**Figure 7.29** Hierarchical data analytics. (A) Orange data analytics framework has many drag and drop widgets for analytics, (B) Image clustering is done using the Hierarchical clustering widget, herein, hierarchical data analytics is used to classify the similarity between the of 36 SEM images of polymer electrolyte fuel cells (PEFC).
*Courtesy: Nissan ARC, Japan.*

the advanced analytics. Alternatively, paid and free software such as MATLAB and ImageJ are also widely used. Ironically, paid software is quite expensive and to use MATLAB one must be proficient in programming. In addition, most of the techniques are only limited to the individual image, and fail to address the issue of a large image data set analysis: FIB sliced 2D image data set from a single sample, FIB sliced multiple 2D image data set from the different data sets. However, the researcher is turning to the machine learning (ML) and artificial intelligence (AI) for image analytics.

Machine learning can be used for image analytics for classification, clustering, and prediction. Machine learning (ML) follows the following standard process: data consolidation (Image collection), feature space extraction (feature extraction and feature selection), model building using the selected features, validating the model, and testing the model. Machine learning can automate the process of feature extraction using some of the advanced analytical techniques such as principal component analysis (PCA), multiple component regression (MCR-ALS), nonnegative matrix factorization (NNMF), artificial neural network (ANN), etc. Although there are many open source libraries that are available for image analytics, nevertheless, we recommend to use the open source software called Orange [27], which provides widgets to import the image data set, image embedding algorithms, and an image viewer for quantitative and qualitative data analytics. Orange is a hassle-free software with just plug-and-play widgets for image analytics as shown in Fig. 7.29.

## 7.7 Conclusions

TEM, SEM, AFM, and the confocal microscope can be used individually to investigate the sample and its various properties. Each technique has its own pros and cons, and at times they complement each other. Often a single technique alone can't provide all the information needed to understand the property of the desired sample. The confocal microscope helps to study the living biological cells or tissues in vitro and even can be used to study the drug response to animal cells using fluorescence calibration. TEM can be utilized to magnify the objects up to 10,000,000 times, to provide access to atomic level internal and cross-sectional volumetric information of biological cells or nanomaterials. SEM can be used to analyze size distribution and the surface/subsurface morphologies of a wide range of samples, both biological and nanoparticle. High-resolution AFM provides surface information and mechanical properties of soft and hard nanomaterials and conformal studies of animal cells.

## Acknowledgment

VK is thankful to Amity University Noida and SERB, India. Ankur Baliyan would like to thank NISSAN ARC LTD & Bio-Nano Electronics Research Center, Toyo University, Japan for their support.

## References

[1] Williams DB, Carter CB. Transmission electron microscopy: a textbook for materials science. 2nd ed. Springer; 2009. Available from: https://doi.org/10.1007/978-0-387-76501-3.

[2] Zou X, Hovmöller S, Oleynikov P. Electron crystallography. Electron. Microscopy and electron diffraction. Oxford University Press; 2012. Available from: https://doi.org/10.1093/acprof:oso/9780199580200.001.0001.

[3] Schrand AM, Schlager JJ, Dai L, Hussain SM. Preparation of cells for assessing ultrastructural localization of nanoparticles with transmission electron microscopy. Nat Protoc 2010;5(4):744−57. Available from: https://doi.org/10.1038/nprot.2010.2.

[4] McIntosh R, Nicastro D, Mastronarde D. New views of cells in 3D: an introduction to electron tomography. Trends Cell Biol 2005;15(1):43−51. Available

[5] De Carlo S, Harris JR. Negative staining and Cryo-negative staining of macromolecules and viruses for TEM. Micron 2011;42(2):117−31. Available from: https://doi.org/10.1016/j.micron.2010.06.003.

[6] https://www.sigmaaldrich.com/labware/labware-products.html?TablePage = 21721280.

[7] Narayan K, Subramaniam S. Focused ion beams in biology. Nat Methods 2015;12(11):1021−31. Available from: https://doi.org/10.1038/nmeth.3623.

[8] https://www.tedpella.com/grids_html/4510half.htm.

[9] Baliyan A, Nakajima Y, Fukuda T, Uchida T, Hanajiri T, Maekawa T. Synthesis of an ultradense forest of vertically aligned triple-walled carbon nanotubes of uniform diameter and length using hollow catalytic nanoparticles. J Am Chem Soc 2014;136(3):1047−53.

[10] Zhang L, Lei D, Smith JM, Zhang M, Tong H, Zhang X, et al. Three-dimensional structural dynamics and fluctuations of DNA-nanogold conjugates by individual-particle electron tomography. Nat Commun 2016;7(11083). Available from: https://doi.org/10.1038/ncomms11083.

[11] Goldstein JI, Newbury DE, Echlin P, Joy DC, Fiori C, Lifshin E. Scanning electron microscopy and X-ray microanalysis: a text for biologists, materials scientists, and geologists. Boston, MA: Springer; 2009. Available from: https://doi.org/10.1007/978-1-4613-0491-3.

[12] Amelinckx S, van Dyck D, van Landuyt J, van Tendeloo G. Electron microscopy: principles and fundamentals. Wiley; 2007. Available from: https://doi.org/10.1002/9783527614561.

[13] https://www.hitachi-hightech.com/eu/product_list/?ld = sms2&md = sms2-1&version =.

[14] Li Y, Yuan H, von dem Bussche A, Creighton M, Hurt RH, Kane AB, et al. Graphene microsheets enter cells through spontaneous membrane penetration at edge asperities and corner sites. PNAS 2013;110 (30):12295−300.

[15] Kim KH, Akase Z, Suzuki T, Shindo D. Charging effects on SEM/SIM contrast of metal/insulator system in various metallic coating conditions. Mater Transact 2010;51(6):1080−3. Available from: https://doi.org/10.2320/matertrans.M2010034.

[16] Baliyan A, Hayasaki Y, Fukuda T, Uchida T, Nakajima Y, Hanajiri T, et al. Precise control of the number of walls of carbon nanotubes of a uniform internal diameter. J Phys Chem C 2013;117(1):683−6. Available from: https://doi.org/10.1021/jp309894s.

[17] Narayan K, Subramaniam S. Focused ion beams in biology. Nat Methods 2015;12(11):1021−31.

Available from: https://doi.org/10.1038/NMETH.3623.

[18] Bonmell D. Scanning probe microscopy and spectroscopy: theory, technique, and application. New York: Wiley-CVH; 2001.

[19] https://afm.oxinst.com/products/cypher-afm-systems/cypher-s-afm.

[20] https://afmprobes.asylumresearch.jp/.

[21] Haga H, Nagayama M, Kawabata K, Ito E, Ushiki T, Sambongi T. Time−lapse viscoelastic imaging of living fibroblasts using force modulation mode in AFM. J Electron Microsc 2000;49:473−81. Available from: https://doi.org/10.1093/oxfordjournals.jmicro.a023831.

[22] Campbell JF, Tessmer I, Thorp HH, Erie DA. Atomic force microscopy studies of DNA-wrapped carbon nanotube structure and binding to quantum dots. J Am Chem Soc 2008;130(32):10648−55. Available from: https://doi.org/10.1021/ja801720c.

[23] Sharma S, Rasool HI, Palanisamy V, Mathisen C, Schmidt M, Wong DT, et al. Structural-mechanical characterization of nanoparticle exosomes in human saliva, using correlative AFM, FESEM, and force spectroscopy. ACS Nano 2010;4 (4):1921−6. Available from: https://doi.org/10.1021/nn901824n.

[24] Matsumoto B. [chapter 1] Methods in cell biology, cell biological applications of confocal microscopy, vol. 70. USA: Elsevier Science; 2002.

[25] https://www.olympus-lifescience.com/en/microscoperesource/primer/techniques/confocal/confocalintro/.

[26] Baligar P, Kochat V, Arindkar SK, Equbal Z, Mukherjee S, Patel S, et al. Bone marrow stem cell therapy partially ameliorates pathological consequences in livers of mice expressing mutant human α1-antitrypsin. Hepatology 2017;65 (4):1319−35. Available from: https://doi.org/10.1002/hep.29027.

[27] https://orange.biolab.si/.

# Chapter | 8 |

# Principles and applications of flow cytometry

*Marcio Chaim Bajgelman*

*Brazilian Biosciences Laboratory, National Center for Research in Energy and Materials, Campinas, Brazil*

## 8.1 Introduction

Flow cytometry is a powerful tool for cell analysis, that allows morphological characterizations such as cell size, granulation, expression of molecular targets, DNA or RNA content, cell counting, and determination of viability, among others. Some equipment also are able to separate cell populations in the process known as cell sorting.

The principle of cytometer functioning consists of making the cells pass through a focused light source in a capillary. Photodetectors capture signals that allow to evaluate cell size, granularity, and expression of molecular targets. These signals from individual cells are processed and analyzed graphically, such as histograms or dot plots, to identify and quantify cell populations, or even to assess target gene expression. Extracellular proteins may also be complexed with appropriate particles and labeled with fluorescent probes or antibodies, making possible their quantification.

### 8.1.1 Light scattering and fluorescence

Light scattering is deflection of incident light by a particle. This phenomenon depends on the physical properties of the particle, such as its size and complexity. Flow cytometry has two different light scattering detectors, the forward scatter (FSC) and the side scatter (SSC). In Fig. 8.1 we can see a schematic of the flow cell, represented by the fluidic system that aligns cells to pass through a light beam. The cell scatters light in different angles or even can be excited to emit fluorescence. Light is collected by photodetectors.

The FSC detector measures the light intensity in the direction of the optical path of the incident source in the direction of the sample. The measured intensity is proportional to the cell diameter, as a function of its diffraction. The FSC parameter allows to compare cell sizes. A critical point that may compromise the analysis of the FSC parameter is the ratio of the cell size to the wavelength of the light source. Particles with a diameter less than wavelength may exhibit altered behavior, causing inconsistency in FSC reading.

The SSC detector measures refracted or reflected light, related to cell granularity and complexity. The SSC signal is of lower intensity than the FSC signal, requiring the use of a photomultiplier for its amplification.

In this way, the FSC and SSC signals allow to morphologically characterize different cellular subpopulations in a sample.

Considering the fluorescence of the sample, it is possible to detect emissions at different wavelengths, allowing a multiple analysis, using a set of lasers and filters for excitation and detection at different wavelengths.

### 8.1.2 Cell labeling and compensation

We can use different strategies for labeling cells for flow cytometry. Cells can be transfected with vectors encoding fluorescent proteins such as GFP [1], or even stained with dyes that impregnate the cell surface as CFSE [2]. Cells can also be permeabilized for direct labeling of their genetic material. Extracellular proteins and intracellular proteins can also be labeled with fluorophores-labeled probes or antibodies [3,4].

Depending on the configuration of the cytometer it is possible to simultaneously analyze different markers,

Data Processing Handbook for Complex Biological Data Sources. DOI: https://doi.org/10.1016/B978-0-12-816548-5.00008-3

depending on the combination and compatibility of fluorophores. It is possible to use different sets of lasers and filters for excitation and emission capture. In addition to the optical assembly, the compensation is fundamental to enable the simultaneous analysis of different fluorophores. Some fluorophores have a broader spectrum of emission and can spill over other channels. Compensation algorithms allow to reduce sample interference on unwanted channels.

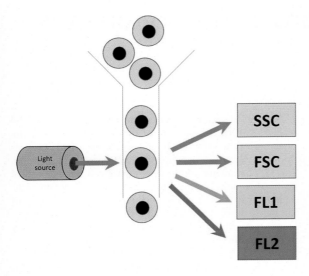

**Figure 8.1** Overview of flow cytometry. Cells pass through a focused light beam. The scattered light is collected by FSC and SSC detectors. Fluorescence is collected by FL1 and FL2 detectors.

## 8.2 Integrating data approaches and analytical tools

Flow cytometry allows a simultaneous analysis of several parameters, depending on limitations and configurations of the equipment, as the availability of different wavelength light source and/or filters for fluorescence detection. Gating is the most versatile tool of cytometry and can integrate several data sets. In this way, we can acquire a sample performing a morphological analysis using a dot plot and select a desired population. This gated population can be further evaluated in a second graph to profile fluorescence. From this acquired data it is also possible to select another population in a second gate to analyze other parameters. Gating limitation is related to hardware configuration and characteristics of fluorescent probes. Next we describe the most used protocols for flow cytometry, with practical examples.

### 8.2.1 Phenotyping cell populations

Flow cytometry is broadly used by clinical laboratories to count blood cells. The leukocytes profile can be performed by flow cytometric immunophenotyping. There are numerous antibodies conjugated to fluorophores that can be used for immunophenotyping by flow cytometry. Combinations of lasers and filters and compensation technologies enable the simultaneous identification of different cell markers, making flow cytometry a powerful analytical tool. In Fig. 8.2A (left panel), we can identify different cell populations from a blood sample, analyzing a FSC × SSC dot plot. The smaller and less complex

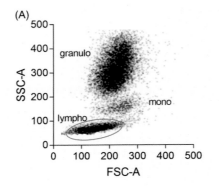

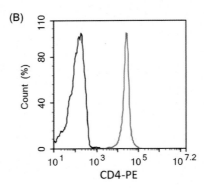

**Figure 8.2** Identification of cell populations by flow cytometry. Left panel shows a FSC × SSC dot plot, indicating three representative populations: lymphocytes, monocytes, and granulocytes. Lymphocyte population is gated in red. Right panels shows an overlay histogram with CD4 T cells stained with an isotype control antibody (*black curve*) and a PE labeled CD4 antibody (*red curve*).

population represented by lymphocytes is shown in the red gate in the lower part of the dot plot, while the granulocytes, with greater internal complexity, are located in the upper part of the graph and the monocytes, of intermediate size and complexity, in the central part. Therefore, SSC and FSC analysis allows for morphological discrimination/quantitation between lymphocytes, monocytes, and granulocytes, but does not allow the differentiation of cellular subtypes, such as CD4 or CD8 T cell subtypes. This kind of profiling, to distinguish morphologically similar subtypes, can be made by using fluorescent markers as antibodies, which will characterize the presence of specific targets that identify cells. Observing the right panel of Fig. 8.2B we can identify CD4 + T cells, gated from the lymphocyte population as indicated in the left panel, that were stained with a PE labeled anti CD4 antibody. In Fig. 8.3, we can observe results of a cytometry analysis of murine regulatory T cells (Treg). Like other T cells, activated Tregs have constitutive expression of surface markers such as 4-1BB, OX40, CTLA-4, among others. Thus, to differentiate Treg from other T cells, it is possible to label the FoxP3 transcription factor, which is considered a master key in the regulation of Treg immunosuppressive phenotype [5,6]. In this way, we labeled cells with an antibody 4-1BB-PE and with an antibody FoxP3-APC, observing the double staining by flow cytometry, which characterizes Treg phenotype. In addition to primary T cells, genetically modified cells can also be characterized by flow cytometry. In Fig. 8.4 we observed a reporter cell line derived from murine melanoma tumor cell B16. These cells show expression of the GFP and the 4-1BB ligand. Labeling this GFP-expressing lineage with an anti-4-1BBL antibody coupled to the phycoerythrin (PE) fluorophore enables simultaneous detection of both transgenes, GFP and 4-1BBL [7].

## 8.2.2 Cell proliferation

Flow cytometry can be used to measure cell proliferation. The most common protocol is to stain target cells with the fluorescent dye carboxyfluorescein succinimidyl ester (CFSE). This dye is covalently coupled to the cell and can be retained for a long period. Due to their high stability, the fluorophore can be observed in the daughter cells, however, reducing intensity to each cell division. In this way, it is possible to monitor the cell proliferation of CFSE-labeled cells. In Fig. 8.5A, right panel, we can see CD4 positive, nonstimulated, T lymphocytes labeled with CFSE. The signal is a sharp peak in the histogram. Stimulating these T cells in the presence of CD3 and CD28 antibodies induces cell proliferation, characterized by the appearance of five additional peaks that exhibit different intensities of CFSE, as a function of cell division.

## 8.2.3 Cell cycle

During the cell cycle there are alterations in the genetic content, which can be observed by fluorescent dye tags, such as propidium iodide. Cell cycle analysis is one of the pioneering applications of flow cytometry. In addition to detecting cell populations at different stages of the cell cycle, such as G0/G1, S, and G2/M, flow cytometry also enables the identification of apoptotic cells (sub G0). In Fig. 8.6 we can see B16 murine melanoma cells that have undergone drug treatment and viral vector transduction for p19 overexpression, as indicated [8]. Cells were fixed after treatment and labeled with propidium iodide to perform cell cycle analysis by a flow cytometry experiment.

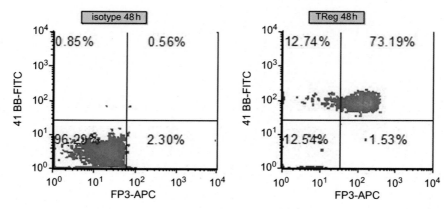

**Figure 8.3** Flow cytometric immunophenotyping of murine regulatory T cells. Cells were stained with a FITC anti 4-1BB MAb and with an APC anti FoxP3 Mab. Left panel shows isotype controls.

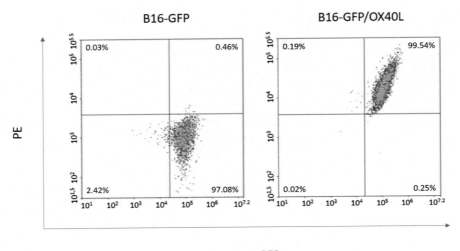

**Figure 8.4** Characterization of genetically modified B16 cells. Cells were transduced with a retroviral vector harboring GFP, establishing the cell line B16-GFP (left panel). The cell line B16-GFP was then transduced with a second retroviral vector encoding OX40 ligand, generating the cell line B16-GFP/OX40L (right panel).
*Adapted from Manrique-Rincon A.J., de Carvalho A.C., Ribeiro de Camargo M.E., Franchini K.G., Bajgelman M.C. Development of a flow cytometry assay which allows to evaluate the efficiency of immunomodulatory vaccines to enhance T cell-mediated antitumor response. J Biotechnol 2018;284:11−16.*

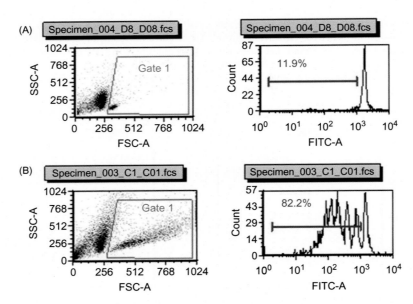

**Figure 8.5** T cell proliferation assay. Murine CD4-Tcells were stained with CFSE. (A) Unstimulated cells. (B) Anti-CD3 and APC stimulation for 96 h.

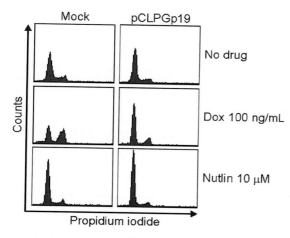

**Figure 8.6** Flow cytometry-based cell cycle assay. B16 murine melanoma-derived cells were transduced with a retrovirus for p19 overexpression (pCLPGp19) and treated with nutlin or doxorubicin, as indicated. Cells were fixed, permeabilized, and stained with propidium iodide to perform flow cytometry.
*Adapted from Merkel C.A., da Silva Soares R.B., de Carvalho A.C., Zanatta D.B., Bajgelman M.C., Fratini P., et al. Activation of endogenous p53 by combined p19Arf gene transfer and nutlin-3 drug treatment modalities in the murine cell lines B16 and C6. BMC Cancer. 2010;10:316.*

### 8.2.4 Apoptosis

As we have mentioned, cell cycle protocol can reveal sub G1 cells, suggesting some process of cell death. This killing process can best be understood by using more refined technologies such as double labeling with propidium iodide (PI) and a fluorescent anti-Annexin V antibody [9]. In Fig. 8.7, left panel, we can observe lymphocytes cultured under normal conditions. These live cells that preserve membrane integrity are not labeled neither by propidium iodide nor antiannexin antibody. When inducing apoptosis with dexamethasone, as shown in the right panel, we observed that cells were labeled with anti-Annexin V antibody, suggesting phosphatidylserine exposure, characteristic of apoptosis. Cells in an advanced process of death also become more permeable, and can be stained by PI.

## 8.3 Conclusion

Flow cytometry is a robust technology for the phenotypic characterization of cells and the understanding of mechanisms and cellular signaling. In this chapter we briefly discussed popular methodologies of flow cytometry, which are employed in research laboratories, biotechnology, and also for diagnostic purposes. It's possible to

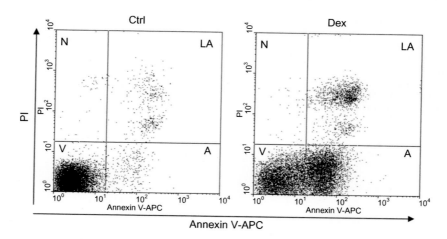

**Figure 8.7** Apoptosis assay. Linfocytes are stained with propidium iodide (PI) and labeled with an anti-Annexin V−FITC antibody. Left panel shows control cells and right panel shows dexamethasone-treated lymphocytes. (V) Live cells: does not stain PI and does not bind annexin, (A) early stage of apoptosis: does not stain PI but binds annexin, (LA) late stage of apoptosis: stains PI and binds annexin, (N) necrotic cells: stains PI but does not bind Annexin.
*Adapted from Wlodkowic D., Skommer J., Darzynkiewicz Z. Flow cytometry-based apoptosis detection. Methods Mol Biol. 2009;559:19−32.*

perform multiple analysis using fluorescent probes/antibodies, depending on compatibility of fluorophores, which are associated to the configuration of the equipment and availability of different wavelength light sources and filters for detection. The cytometry equipment usually has its own software for data acquisition and processing, but there is other compatible software that provides robust platforms for data analysis, like FCS Express and FlowJo, which provide very intuitive platforms and flexibility for analysis and data exportation.

The cytometry equipment has experienced great advances, reducing size of the machine, reducing reagent consumption, and using friendlier and more accessible acquisition and analysis platforms. Some of these devices already allow capturing images of representative cells of a population, combining advantages of fluorescence microscopy and the robustness of flow cytometry. Cytometry can also be used in conjunction with the next generation sequencing, enabling applications in diagnosis and development of new therapeutic strategies [10,11].

# References

[1] Shimomura O, Johnson FH, Saiga Y. Extraction, purification and properties of aequorin, a bioluminescent protein from the luminous hydromedusan, Aequorea. J Cell Comp Physiol 1962;59:223−39.

[2] Lyons AB, Parish CR. Determination of lymphocyte division by flow cytometry. J Immunol Methods 1994;171(1):131−7.

[3] de Vries E, Noordzij JG, Kuijpers TW, van Dongen JJ. Flow cytometric immunophenotyping in the diagnosis and follow-up of immunodeficient children. Eur J Pediatr 2001;160(10):583−91.

[4] Millard I, Degrave E, Philippe M, Gala JL. Detection of intracellular antigens by flow cytometry: comparison of two chemical methods and microwave heating. Clin Chem 1998;44(11):2320−30.

[5] Fontenot JD, Gavin MA, Rudensky AY. Foxp3 programs the development and function of CD4 + CD25 + regulatory T cells. Nat Immunol 2003;4(4):330−6.

[6] Hori S, Nomura T, Sakaguchi S. Control of regulatory T cell development by the transcription factor Foxp3. Science 2003;299(5609):1057−61.

[7] Manrique-Rincon AJ, de Carvalho AC, Ribeiro de Camargo ME, Franchini KG, Bajgelman MC. Development of a flow cytometry assay which allows to evaluate the efficiency of immunomodulatory vaccines to enhance T cell-mediated antitumor response. J Biotechnol 2018;284:11−16.

[8] Merkel CA, da Silva Soares RB, de Carvalho AC, Zanatta DB, Bajgelman MC, Fratini P, et al. Activation of endogenous p53 by combined p19Arf gene transfer and nutlin-3 drug treatment modalities in the murine cell lines B16 and C6. BMC Cancer 2010;10:316.

[9] Wlodkowic D, Skommer J, Darzynkiewicz Z. Flow cytometry-based apoptosis detection. Methods Mol Biol 2009;559:19−32.

[10] Getta BM, Devlin SM, Levine RL, Arcila ME, Mohanty AS, Zehir A, et al. Multicolor flow cytometry and multigene next-generation sequencing are complementary and highly predictive for relapse in acute myeloid leukemia after allogeneic transplantation. Biol Blood Marrow Transplant 2017;23(7):1064−71.

[11] Picot T, Aanei CM, Flandrin Gresta P, Noyel P, Tondeur S, Tavernier Tardy E, et al. Evaluation by flow cytometry of mature monocyte subpopulations for the diagnosis and follow-up of chronic myelomonocytic leukemia. Front Oncol 2018;8:109.

# Chapter | 9 |

# Isothermal titration calorimetry

*Vijay Kumar Srivastava[1],* and Rupali Yadav[2],**

[1]*Amity Institute of Biotechnology, Amity University Rajasthan, Jaipur, India,* [2]*Dr. Reddy's Institute of Life Sciences, University of Hyderabad Campus, Hyderabad, India*

## 9.1    Introduction

Isothermal titration calorimetry (ITC) is one of the physical techniques that directly measures the heat discharged or consumed all along a bimolecular reaction. It is an analytical method where the ligand comes in contact with a macromolecule under constant temperature [1]. It works on the basic principle of thermodynamics where contact between two molecules results in either heat generation or absorption, depending on the type of binding, that is, exothermic or endothermic [2]. The instrument consists of two cells; one is the main cell for the macromolecule of concern and the other cell is called a reference cell, which is meant for the solvent, as shown in Fig. 9.1.

Both cells are kept at steady temperature and pressure. The ligand is sucked through a syringe and titrated into the main cell. Macromolecular binding with the ligand results either in heat discharge or consumption, which causes the change of temperature within the main cell. However, the instrument will always maintain the constant temperature in the main cell equivalent to that of the reference cell. For maintaining the temperature, the instrument gives relevant power (higher or lower) depending on interactions (Fig. 9.2).

The heat change is then simply calculated by integrating the power over the time (seconds), which gives us the enthalpy of the reaction. The heat discharged or consumed all along the calorimetric reaction corresponds to

the fraction of bound ligand and increased ligand concentration leads to saturation of substrate and finally less heat is discharged or consumed.

The amount of heat discharged upon inclusion of ligand is defined as follows [4]:

$$Q = V_o \, \Delta H_b \, [M]_t \left\{ \frac{K_a \, [L]}{1 + K_a \, [L]} \right\}$$

where $Q$ is heat evolved/absorbed; $V_o$ is sample cell volume; $\Delta H_b$ is enthalpy of binding per mole of ligand; $[M]_t$ corresponds to the total concentration of macromolecule in the sample cell; $K_a$ is binding constant; $[L]$ is concentration of free ligand.

Accurate measurement of released/absorbed heat is then used to determine the binding constants ($K_a$), enthalpy ($\Delta H$), entropy ($\Delta S$), and the reaction stoichiometry ($n$), thereby an entire thermodynamic parameter of the molecular binding can be obtained in an individual analysis by applying this technique (Fig. 9.3).

A very important aspect that determines the success of an isothermal titration calorimetric attempt is deciding the actual concentration for protein and ligand respectively. The appropriate protein concentration relies on the parameter "$c$" value, which is defined as:

$$c = nP_t/K_d$$

where $P_t$ is the concentration of protein kept in the measurement cell, "$n$" is the number of binding sites per protein molecule, and $K_d$ is the affinity constant [5]. The $c$ value determines the architecture of the binding isotherm. The higher $c$ value ($\sim 1000$) will lead to the generation of too-steep shaped curve, which will make it difficult to get an estimate of $K_d$, although other parameters such as $\Delta H$ and $n$ can be determined. If the $c$ value is $<5$, then

---

* Both the authors contributed equally.

architecture of the binding isotherm is too shallow, which does not allow precise determination of the thermodynamic parameters unless one of them is earlier known (such as the stoichiometry). Hence, to get a good sigmoidal shape of the binding isotherm so as to estimate the $K_d$, $\Delta H$, $\Delta S$, and $n$, it is necessary to keep the $c$ value between 20 and 100.

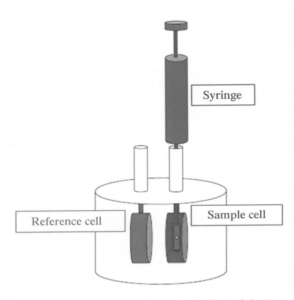

**Figure 9.1** Instrumentation of the typical isothermal titration calorimetry [3].

## 9.2 Raw data

The raw signal in isothermal calorimetry is the power ($\mu$cal s$^{-1}$ or $\mu$J s$^{-1}$) used to control the heater to keep the temperature of cell the same as a function of time as represented in Fig. 9.4. Sample raw data is illustrated in Fig. 9.5.

To further determine the heat produced after every injection, the area under each peak is obtained by using the following equation [7]:

$$q_i = v \times \Delta H \times \Delta L_i$$

which demonstrates that the heat released or absorbed ($q_i$) during each injection of a reaction corresponds to the extent of ligand that interacts with protein in a precise injection ($v \times \Delta L_i$) and the distinctive binding enthalpy ($\Delta H$) for the reaction. The quantity $\Delta L_i$ denotes the concentration difference between the bound ligand in the $i$th and $(i-1)$th injections. This is dependent upon the specific binding model, reflective of whether the interaction follows single, multiple, or cooperative binding model. The most common case is the protein with a single interaction site; the above equation evolves into:

$$q_i = v \times \Delta H \times [P] \times \left( \frac{K_a[L]_i}{1 + K_a[L]_i} - \frac{K_a[L]_{i-1}}{1 + K_a[L]_{i-1}} \right)$$

where $K_a$ and $[L]$ is the binding constant and concentration of free ligand respectively [7].

Therefore, heat after every injection is determined by estimating the area under each peak. Since the amount of free protein (not bound to ligand) decreases

**Figure 9.2** Representative model showing how ITC works [3].

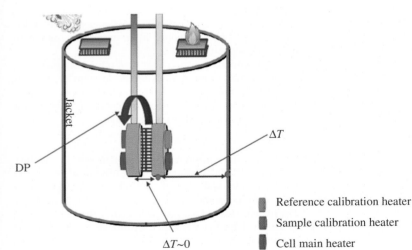

Jacket

DP

$\Delta T$

$\Delta T \sim 0$

■ Reference calibration heater

■ Sample calibration heater

■ Cell main heater

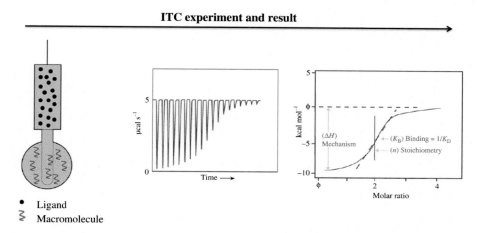

Figure 9.3 Pictorial representation showing the progress of ITC experiment with result [3].

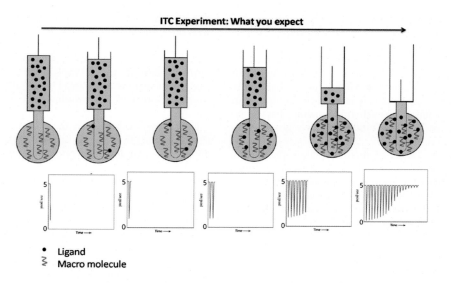

Figure 9.4 Pictorial representation showing the stepwise progress of ITC experiment result.

progressively after each successive injection as the reaction proceeds, accordingly the magnitude of the peak also reduces continuously until saturation is accomplished. Once the saturation is reached, consecutive injections produce the same peaks equivalent to the heat of dilution. To further obtain the heat of binding, the observed binding peaks are integrated and subtracted from the heat of dilution, which is identified by injecting the ligand solution into the buffer.

## 9.3 Data processing

To obtain the thermodynamic parameters, raw data is further processed for baseline correction, subtraction of heat of dilutions using the ITC software. In the subsequent discussion, we will explain all the steps involved in data processing using the Nanoanalyze software. Finally, the processed experimental data is fitted to a particular model

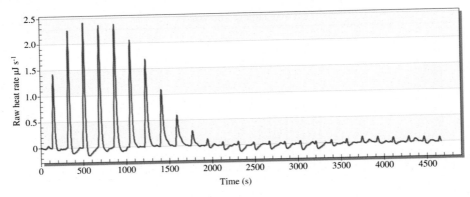

**Figure 9.5** Calorimetric titration of the native vacuolar sorting protein 29 from *Entamoeba histolytica* with divalent metal Zn representing the raw ITC data acquired from 25 injections (1 μL each) of divalent metal that is Zn into 100 μM of protein (*Eh*Vps29) [6].

(independent model, multiple site model, cooperative model, etc.) given in the Nanoanalyze software. It is generally recommended to first fit the data in the independent model if you do not have prior information about the stoichiometry of the interaction. Then try to fit the data in other models as well and check the results. Finally, the correct model for the data is selected by looking at the mathematical analysis for thermodynamic parameters acquired from each of the binding models given in the Nanoanalyze software.

To identify the values of thermodynamic parameters, the thermograms are usually identified by a curve fitting process that utilizes the nonlinear regression analysis so as to obtain the best fit of the model. The analysis gives the values of stoichiometry ($n$), affinity constant ($K$), and enthalpy ($\Delta H$) for the binding reaction. Then, free energy of binding is determined from these values using the equation:

$$\Delta G = -RT\ln K$$

The value of entropy is determined by the following equation:

$$\Delta G = \Delta H - T\Delta S$$

Statistical analysis for thermodynamic parameters can be executed using the program inbuilt in Nanoanalyze software.

## 9.3.1 Web-based sources

There are no freely available automatic software programs for processing the ITC data. These software programs are in-built programs provided by the manufacturer along with the instrument. Although, the data can be processed with any scientific program available for graphing and data analysis such as Origin, Sigma plot, etc. Here for further discussion, we will explain the ITC data processing using Nanoanalyze software. First of all, install the software and open it. The window will look as shown in Fig. 9.6.

Now, open both the raw data, interaction and control (heat dilution) by going to file and open option given in the software window. The raw data appears as shown in Fig. 9.7.

To further process the data it is necessary to correct the baseline of the data. This can be done by baseline correct option given in the software as shown in Fig. 9.7.

After baseline correction we need to subtract the heat of dilution data from the interaction data, which is very important to determine the actual heat of interaction. For this go to the Area option, which will show an option for area correction, as shown in Fig. 9.8.

The heat of dilution must be subtracted. Go to the Area option, drag and drop the heat of dilution data to blank (Figs. 9.9 and 9.10).

We are ready with the data for modeling it into particular model so as to determine the thermodynamic parameters of the interaction after subtracting the heat of dilution. For this purpose, go to the model section in the modeling area, as shown in Fig. 9.11. It will again open a new window showing different models.

It is generally recommended that if you do not have any prior information regarding the interaction, try to first fit the data in the independent model. This model tries to fit the data considering that there is only one binding site.

**Figure 9.6** Snapshot showing the Nanoanalyze software window.

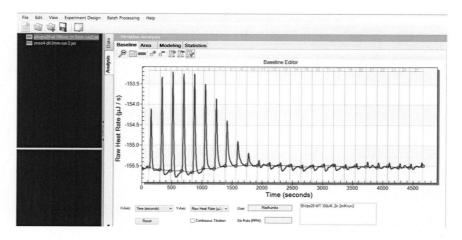

**Figure 9.7** Snapshot showing the raw data in Nanoanalyze software window.

Now, data is ready to fit into independent model. Just start the fitting procedure by clicking on the green button in Fit data option. Data fitting is an iterative process that ultimately gives the list of thermodynamic parameters such as enthalpy ($\Delta H$), entropy ($\Delta S$), affinity constant ($K_d$), and the stoichiometry of the interaction ($n$) as shown in Fig. 9.12.

Now, after processing the data it is very important to check its statistical reliability. Statistical analysis for thermodynamic parameters can be executed using the program provided with the Nanoanalyze software by keeping

the constant value of standard deviation (0.1) and desired confidence level (99%) with 1000 trials as shown in Fig. 9.13.

Finally, when the statistical analysis is complete, it will give the histograms as shown in Fig. 9.14 along with the values of all the thermodynamic parameters with associated standard error. The good Gaussian histogram signifies the good statistics of the data. If this is not good, you can again try to fit the data in different models and check the statistics of the parameters. By comparing the statistics of the thermodynamic

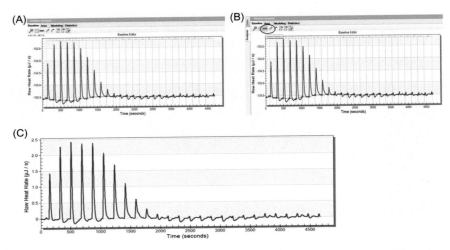

**Figure 9.8** Baseline correction. (A) Raw data without baseline correction. (B) For baseline correction click on the option highlighted with black circle. (C) Data after baseline correction.

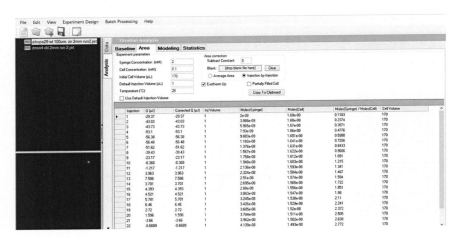

**Figure 9.9** Snapshot showing the window for area correction to subtract the heat of dilution from interaction data.

parameters in different models of the same data, you can confidently know the correct model for your data. As mentioned in the previous section, the statistical analysis for thermodynamic parameters can be executed by keeping the constant value of standard deviation (0.1) and desired confidence level (99%) with 1000 trials. The values of standard deviation, confidence level %, and the number of trials can be changed. Generally, lower value of standard deviation (0.1−1.0), higher confidence level %, and greater number of trials give the expected good Gaussian shape to distributions of the parameters.

## 9.4 Examples

Characterization of thermodynamic interactions between the molecules (proteins, nucleic acids, and lipids) is important to understand the functional aspects of the biological systems. Although, recent advances in the field of structure biology have provided the mechanistic view of the interactions at atomic level with a biochemical and functional correlation, still the picture is not complete and requires detailed characterization of the interactions involving estimation of affinity, number of binding sites

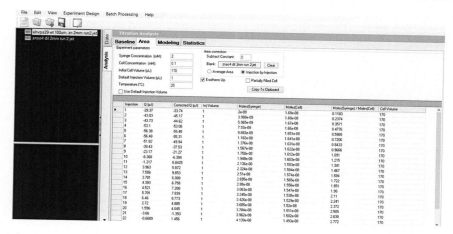

**Figure 9.10** Snapshot showing the window for subtracting the heat of dilution from the interaction data.

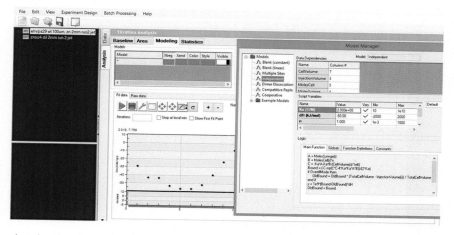

**Figure 9.11** Snapshot showing the window for modeling the data in independent model.

and thermodynamic parameters. This helps in describing the function and mechanism at the molecular level.

The ITC technique is used for the thermodynamic characterization of biomolecules in solution and is universally accepted for studying protein–protein interactions, and protein–ligand interactions. In subsequent discussions, with the help of selected examples we will explain the complete ITC data analysis and its interpretation.

## 9.4.1 Protein–metal interactions

Here we take the example of amoebic protein Vps29 (EhVps29) and its interaction with zinc metal. The choice

of control is very important, that is, heat of ligand dilution as mentioned in previous sections. But apart from this, you need to include some negative control, that is, interaction with any nonspecific ligand in your studies for assessing the specificity of the interaction. In a recent study, the author assessed the metal specificity of EhVps29, comprehensive studies on its interaction with several divalent metals (Mg, Mn, Zn, and Ca) using ITC [6]. Interestingly, among several metals, they identified that the binding of vacuolar sorting protein 29 from *Entamoeba histolytica* is preferable to zinc. Table 9.1 represents the summary of all the thermodynamic parameters obtained for the interaction of EhVp29 to zinc.

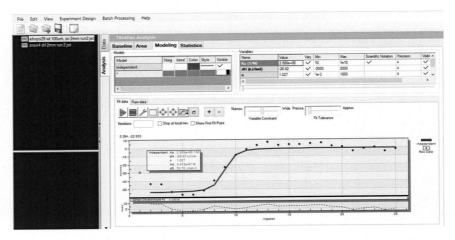

**Figure 9.12** Snapshot showing the window for modeled data in independent model with all the thermodynamic parameters.

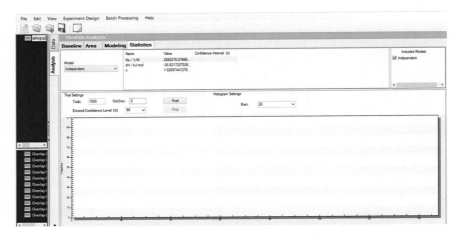

**Figure 9.13** Snapshot showing the window for checking the statistics of the modeled data in independent model for all the thermodynamic parameters.

As shown in Fig. 9.15 and Table 9.1, EhVps29 binds to zinc metal with micromolar affinity $(-2.9 \times 10^6\,\mathrm{M}^{-1})$. Overall $Zn^{++}$ binding to EhVps29 is found to be spontaneous and both the enthalpic as well as entropic factors contributed towards the feasibility of interactions [6].

In the above discussion, you have seen the example of protein–metal interactions with favorable enthalpy and entropy. Now we will introduce another example of protein–metal interaction, which is spontaneous but with favorable enthalpic contributions and unfavorable entropic contribution (decrease in entropy).

### 9.4.2 Thermodynamic basis of calcium binding to EhCRD from *Entamoeba histolytica*

To establish EhCRD as a calcium-binding protein, authors have analyzed its interaction with calcium using ITC [8]. As shown in Fig. 9.16, the affinity and stoichiometry of EhCRD to calcium ion found to be in micromolar range (an association constant of $1.0 \times 10^5\,\mathrm{M}^{-1}$) and one respectively.

Thermodynamic parameters for the calcium binding to EhCRD are given in Table 9.2. These values of

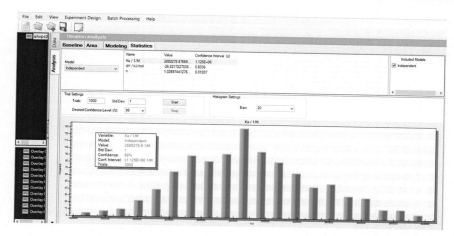

**Figure 9.14** Snapshot showing the result of the statistical analysis of the modeled data in independent model for all the thermodynamic parameters.

**Table 9.1 Thermodynamic parameters for the interaction of wild type EhVps29 to several metals.**

| Protein (100 μm) | Ligand (2 mM) | $\Delta H$ (kJ mol$^{-1}$) | $\Delta S$ (J mol$^{-1}$ K$^{-1}$) | $T\Delta S$ (kJ mol$^{-1}$) | $\Delta G$ (kJ mol$^{-1}$) | $K_a$ (M$^{-1}$) |
|---|---|---|---|---|---|---|
| EhVps29 wild type | ZnSO$_4$ | −26.8 | 33.7 | 10.1 | −36.8 | $2.9 \times 10^6$ |
| EhVps29 wild type | MnCl$_2$ | | | No interaction | | |
| EhVps29 wild type | MgCl$_2$ | | | No interaction | | |
| EhVps29 wild type | CaCl$_2$ | | | No interaction | | |

thermodynamic parameters suggest that the interaction is spontaneous, which is majorly driven by enthalpy with reduction in entropy. Entropic reduction is attributed to loss in conformational degree of freedom of the protein upon binding to calcium ion. Further, to rule out the possibility of its interaction with other metals, authors have also included Mg$^{++}$ ion in their study. As a result, they found that EhCRD does not bind to magnesium. This further confirms the metal specificity of EhCRD for calcium [8].

### 9.4.3 Protein carbohydrate interactions

Here, we are taking the example of EhCRD, that is, carbohydrate recognition domain of Gal/GalNAc lectin from *E.*

*histolytica* and its interaction with synthetic sugar, Gal-GalNAc [8]. Table 9.3 represents the thermodynamic parameters for interaction of EhCRD to Gal-GalNAC sugar. EhCRD binds to Gal-GalNAc with micromolar affinity (association constant of $2.3 \times 10^6$ M$^{-1}$).

To assess the sugar specificity of EhCRD, the authors have also studied its interaction with maltose as a negative control. As shown in Fig. 9.17, EhCRD did not interact with maltose. Binding was spontaneous and favored by entropic contribution with unfavorable enthalpy. The observed positive enthalpy and entropy for Gal-GalNAc interaction with EhCRD might be because of reorganization of water molecules resulting from the release of structured water molecules upon binding. Literature reports substantiating these findings are not rare. Multiple investigations on multivalent

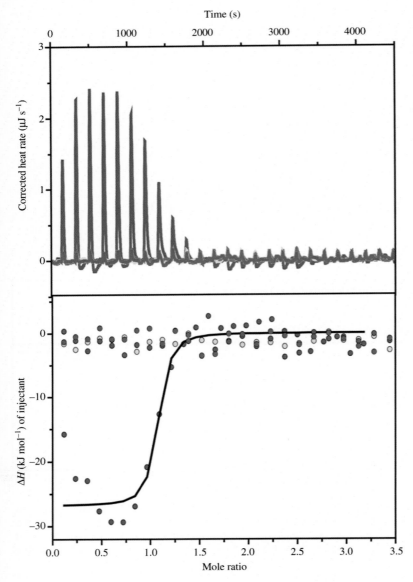

Figure 9.15 Interaction of vacuolar sorting protein 29 form *E. histolytica* with several metals. Upper panel shows the raw data for the binding of EhVps29 to $Zn^{++}$ (*blue*), $Mn^{++}$ (*red*), $Mg^{++}$ (*green*), and $Ca^{++}$ (*yellow*). The lower panel represents the integrated peaks obtained from the raw data for the interaction of EhVps29 with several metals. EhVps29 showed interaction with zinc metal. The other tested metals ($Mn^{++}$, $Mg^{++}$, and $Ca^{++}$) did not show any binding [6].

lectin−carbohydrate interaction reported the importance of positive entropic contributions in stabilizing, making the interaction more feasible and for the increase in affinity for the multivalent carbohydrate compared to monovalent analogs [9−11]. The stoichiometry of EhCRD for the Gal-GalNAc was found to be 0.5. Since EhCRD exists as a dimer in vitro, the fractional value of stoichiometry (0.5) raises the possibility of cross-linking of protein upon carbohydrate binding. Similar phenomenon has also been documented for protein Concanavalin A [12]. The fractional values for stoichiometry ($n$) for lectin−carbohydrate interactions signify the carbohydrate mediated cross-linking in lectins [13]. Hence, it was proposed that the fractional stoichiometry could be due to the binding of Gal-GalNAc to a cross linked EhCRD.

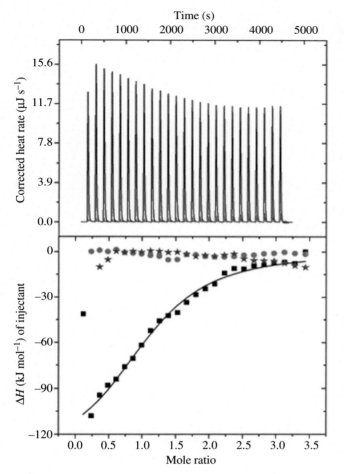

Figure 9.16 Interaction of EhCRD with calcium. Upper panel represents the raw ITC data for the binding of *Eh*CRD to calcium (*black square*) and MgCl$_2$ (*red circle*). The lower panel represents the integrated normalized data fitted into independent model of binding [8].

**Table 9.2  Thermodynamic parameters of binding of EhCRD wild type with calcium.**

| Protein | Ligand | [EhCRD] ($\mu$m) | [Ligand] (mM) | $\Delta H$ (kJ mol$^{-1}$) | $\Delta S$ (J mol$^{-1}$ K$^{-1}$) | $T\Delta S$ (kJ mol$^{-1}$) | $\Delta G$ (kJ mol$^{-1}$) | $K_a$ (M$^{-1}$) |
|---|---|---|---|---|---|---|---|---|
| EhCRD wild type | CaCl$_2$ | 50 | 0.5 | −136.0 | −362.0 | −108.0 | −27.6 | $1.0 \times 10^5$ |

**Table 9.3  Thermodynamic parameters for the binding of EhCRD to Gal-GalNAC sugar.**

| Protein | Ligand | [EhCRD] ($\mu$m) | [Ligand] (mM) | $\Delta H$ (kJ mol$^{-1}$) | $\Delta S$ (J mol$^{-1}$ K$^{-1}$) | $T\Delta S$ (kJ mol$^{-1}$) | $\Delta G$ (kJ mol$^{-1}$) | $K_a$ (M$^{-1}$) |
|---|---|---|---|---|---|---|---|---|
| EhCRD wild type | Gal-GalNAc | 25 | 0.1 | 933.0 | 3251.0 | 969.0 | −36.0 | $2.3 \times 10^6$ |

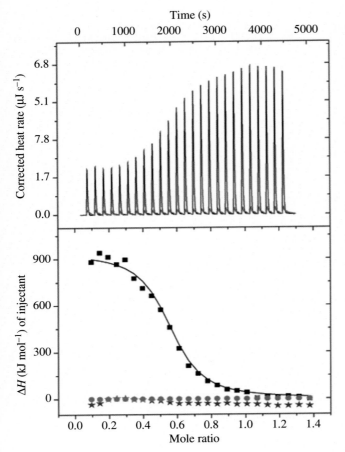

**Figure 9.17** Interaction of EhCRD to Gal-GalNAc sugar. Raw data for the interaction of EhCRD to Gal-GalNAc (*black square*) and maltose (*red circle*) is shown in the top panel. The lower panel represents the integrated normalized data fitted into independent model of binding [8].

## 9.5 Conclusion

The chapter contains the basic knowledge about ITC and how a layperson can perform the experiment to identify the interactions between the two molecules. We have explained the experimental procedure with the help of suitable examples exhibiting exothermic and endothermic reactions, describing in detail the raw data processing to obtain the various thermodynamic parameters. We sincerely hope our efforts will be embraced by students with appreciation and enthusiasm for learning.

## References

[1] Wiseman T, Williston S, Brandts JF, Lin LN. Rapid measurement of binding constants and heats of binding using a new titration calorimeter. Anal Chem 1989;179(1):131−7.

[2] Indvk L, Fisher HF. Theoretical aspects of isothermal titration calorimetry. Methods Enzymol 1998;64:295−350.

[3] http://www.imb.sinica.edu.tw/en/research/core_facilities/biophysicals/form/Biophysics_iTC200.pdf.

[4] Leavitt S, Freire E. Direct measurement of protein binding energetics by isothermal titration calorimetry. Curr Opin Struct Biol 2001;11:560−6.

[5] Dutta AK, Rösgen J, Rajarathnam K. Using isothermal titration calorimetry to determine thermodynamic parameters of

protein—glycosaminoglycan interactions. Methods Mol Biol 2015;1229:315—24.

[6] Srivastava VK, Yadav R, Natsuki W, Tomar P, Mukherjee M, Gourinath S, et al. Structural and thermodynamic characterization of metal binding in Vps29 from *Entamoeba histolytica*: implication in retromer function. Mol Microbiol 2017;106 (4):562—81.

[7] Freyer MW, Lewis EA. Isothermal titration calorimetry: experimental design, data analysis, and probing macromolecule/ligand binding and kinetic interactions. Methods Cell Biol 2008;84:79—113.

[8] Yadav R, Verma K, Chandra M, Mukherjee M, Datta S. Biophysical studies on calcium and carbohydrate binding to carbohydrate

recognition domain of Gal/GalNAc lectin from *Entamoeba histolytica*: insights into host cell adhesion. J Biochem 2016;160(3):177—86.

[9] Dam TK, Roy R, Das SK, Oscarson S, Brewer CF. Binding of multivalent carbohydrates to concanavalin A and *Dioclea grandiflora* lectin. Thermodynamic analysis of the "multivalency effect. J Biol Chem 2000;275(19):14223—30.

[10] Dam TK, Oscarson S, Brewer CF. Thermodynamics of binding of the core trimannoside of asparagine-linked carbohydrates and deoxy analogs to *Dioclea grandiflora* lectin. J Biol Chem 1998;273 (49):32812—17.

[11] Dam TK, Gerken TA, Brewer CF. Thermodynamics of multivalent carbohydrate-lectin cross-linking

interactions: importance of entropy in the bind and jump mechanism. Biochemistry 2009;48 (18):3822—7.

[12] Mandal DK, Kishore N, Brewer CF. Thermodynamics of lectin-carbohydrate interactions. Titration microcalorimetry measurements of the binding of N-linked carbohydrates and ovalbumin to concanavalin A. Biochemistry 1994;33 (5):1149—56.

[13] Dam TK, Gerken TA, Cavada BS, Nascimento KS, Moura TR, Brewer CF. Binding studies of alpha-GalNAc-specific lectins to the alpha-GalNAc (Tn-antigen) form of porcine submaxillary mucin and its smaller fragments. J Biol Chem 2007;282(38):28256—63.

# Chapter | 10 |

# Metagenome analysis and interpretation

*Nar Singh Chauhan*

*Department of Biochemistry, Maharshi Dayanand University, Rohtak, India*

## 10.1   Introduction

Microbes have been identified in all possible habitats. The expedition to uncover the microbial world started with the discovery of microscope. Until the 1990s, a number of microbes were cultured from various habitats to define their structure, physiology, and their possible function as a part of that ecosystem. However, in 1985, Staley and Konopka coined "the great plate count anomaly" describing that the number of cells observed with a microscope is far higher than the number of colonies appearing on agar medium and indicating the presence of unculturable bacteria in any habitat [1]. The concept of unculturable bacteria was proven in 1995 when Amann et al. performed direct sequencing of SSU rRNA gene from environmental DNA to describe microbial diversity of a sample [2]. This study identified a large number of microbial signatures that were not discovered until that date. These microbes have represented >90% of the total microbial diversity of any environment [2,3]. These uncultured microbial groups possess a huge percent of the untapped gene pool that could harbor a number of novel genes or pathways for various biotechnological applications [4,5]. Hereby, a number of efforts were made to cultivate these yet-to-be cultured or uncultured microbes. Despite enormous efforts very limited success was achieved [6]. It was getting difficult to access the gene pool of these microbes and accordingly a number of culture-independent methods were developed that could unveil the untapped gene pool of uncultured microbes [3]. Metagenomics was identified as a pioneer culture-independent approach to gain access to physiology and genetics of microorganisms growing in

any habitat [3,7]. Metagenomics has also been described as environmental genomics, which involves direct extraction of DNA from any environment followed by genomic elucidation either through cloning environmental DNA into a suitable vector followed by its transformation into a culturable host like *Escherichia coli* to identify a gene of interest (i.e., functional metagenomics) [7,8] or sequencing of a target gene or total environment DNA to identify genes/pathways (i.e., sequence metagenomics) [7,8]. In the late 20th century, sequencing was performed mainly through capillary sequencing methods, which were a costly process, hence functional metagenomics gained popularity and a number of novel genes/pathways were identified from various environments using this metagenomics approach. Sequencing driven metagenomics was limited to a few SSU rRNA gene-based microbial diversity studies. Sequencing data output from these studies was limited and generally processed manually [9]. In the early 21st century, just after completion of the Human Genome Project, a number of next-generation sequencing techniques paved their way into research institutions. These sequencing platforms allowed to execute a number of metagenomic sequencing projects (the Sargasso Sea, acid mine drainage, etc.) with their promise of "more data and low cost" [10]. Later on, these next-generation sequencing platforms were elements of choice for metagenomic studies. Usage of next-generation sequencing platforms has allowed researchers to generate huge metagenomic datasets. However, due to nonavailability of appropriate software or pipelines, it became a headache to analyze these datasets to decode microbiome structure and functions. Accordingly, a number of pipelines were discovered to fill this gap [11–13]. The current chapter has been designed to inform the readers about basics of metagenomics and various methods/tools to analyze these metagenome datasets.

## 10.2 Metagenomics

The biosphere is dominated by microorganisms, yet a majority of microbes in nature have not been studied. To date, only a small fraction of the global microbial diversity has been described in the literature. It has become clear that the majority of bacteria in environmental samples remain unculturable. Most of the current approaches used for culture dependent exploration of microbial diversity are biased because of the limitation of cultivation methods that result in the loss of a major portion of the microbial communities (>99%) [14]. To explore unculturable members of the microbial communities, a culture-independent method was designed to access the gene pool of these unculturable bacteria [15]. Metagenomics, the genomic analysis of a population of microorganisms, has emerged as a powerful tool, which gives a better understanding of the community structure and their physiological importance in the ecosystem. Metagenomics is a culture-independent approach to gain access to physiology and genetics of microorganisms growing in an ecosystem. Metagenomics has become a powerful tool to study microbial community of any environment without culturing them in the laboratory. Metagenomics has also been described as a community or environmental genomics that involves direct extraction of DNA from an environment such as soil, water, gut microbiome, etc. This approach is based on the direct extraction of nucleic acids from any environment, cloning into suitable vector and transformation into a culturable host organism. In 1985, Pace and his colleagues introduced an idea of extraction and cloning of DNA directly from an environmental sample [16], but this approach was used first time in 1991 by Schmidt and his coworkers. They isolated the DNA from picoplankton sample and cloned in a phage vector for genomic analysis [17]. This extracted DNA contains pooled genomic DNA from all assemblage of microorganisms present on a particular environmental niche, so-called metagenomic DNA [18,19] and the term *metagenome* was coined by Jo Handlesman in 1998 [20]. Thus, metagenomic DNA represents the whole genetic contents of available microorganisms that could be explored directly without culturing them. Isolated metagenomic DNA from an environmental sample is partially digested and cloned into a suitable vector such as the plasmid, cosmid, fosmid, or bacterial artificial chromosome (BAC), and then transformed into a suitable host

(most preferred *E. coli*) to construct the metagenomic library. These metagenomic libraries from different environmental samples can be screened for various novel genes or pathways in two ways: (1) functional metagenomics, that is, screening on the basis of particular function; or (2) sequence-based metagenomics, that is, screening by directly sequencing the isolated metagenomic DNA [21].

Functional metagenomics is an approach to screen a metagenomic library for particular genes for a specific function. Many novel genes coding different functions have been identified through cloning and heterologous expression of metagenomic DNA and subsequent phenotypic detection of the desired trait conferred on the cloning host. *E. coli* is the most preferred host organism in metagenomic library construction due to the lack of recombination genes (mcrA, mcrBC, recA, recBC), which is important for cloning of foreign insert DNA into a different type of host *E. coli*. Still, host biasness for gene expression is problematic during expression of foreign genes in *E. coli*. Thus, it is the limitation that most of the genes will not be expressed in any particular host bacterium selected for cloning. Despite these limitations, a number of novel genes encoding various enzymes and bioactive molecules have been identified using the functional metagenomics approach [22–26].

Sequence-based metagenomics is generally PCR based or direct sequencing approach to determine DNA sequences of the whole metagenome. This approach involves the design of DNA primers, derived from conserved regions of already-known genes or protein families followed by their direct amplification from metagenomic DNA and sequencing [27].

## 10.2.1 Metagenomic applications

It has been estimated that >99% of the microbes in nature are nonculturable by available techniques. These unculturable microbes are expected to possess novel genes/pathways because of their known phylogenetic and physiological diversity [28]. Metagenomics is an expanding field within microbial ecology that provides access to the genomes of the total microbial community (including the nonculturable microorganisms) in any given environment. This approach has enriched our understanding of the uncultured microbial world and helped in encountering novel genes. Metagenomics provides an unlimited resource for the development of novel genes, enzymes, natural products, bioactive compounds, and bioprocesses that may substantially impact industrial and biotechnological applications.

### 10.2.1.1 Enzymes and metagenomics

The metagenomic approach has identified novel enzymes and biocatalysts that could play a significant role in the field of biotechnology [29]. Various enzymes produced by metagenomic approach are listed in Table 10.1.

### 10.2.1.2 Metagenomics and bioactive molecules

In the latter half of the 20th century, human health was improved by microbe-derived antibiotics and other bioactive agents. When we isolate these antibiotics and medical agents from microorganisms with culture-dependent approach, then ∼99% of known antibiotics and other products will remain rediscovered. Thus, metagenomics may be the richest source for identification of new antibiotics, antimicrobials, and natural products from unculturable microorganisms of different samples (Table 10.2).

### 10.2.1.3 Novel biosynthetic pathways

Functional and sequence-based metagenomic approaches have deciphered various antibiotic resistant, salt stress resistant, acid resistance pathways, etc., from uncultured microorganisms of various environments. Some novel pathways have been listed in Table 10.3. Metagenomics is the study of genomes recovered directly from environmental samples. This approach is often used to understand the genetic composition and metabolic capacity of microbial communities. Metagenomic technologies are originated from environmental microbiology studies and their wide applications have been greatly facilitated by next-generation high throughput sequencing technologies. With the help of next-generation sequencing technology, we are able to examine the diversity of organisms present in specific environments as well as analyze the complex interactions between members of a specific environment.

## 10.2.2 Metagenome analysis

### 10.2.2.1 In silico analysis of the functional metagenomics datasets

Functional metagenomics leads to the identification of genetically active loci for a functional trait. In most of the cases, functional metagenomics ends with novel genetic fragment sharing either nil or very poor homology in the database [3]. It is always a key challenge to correctly identify and characterize the gene responsible for a functional trait. Accordingly, these genetic loci need to be characterized for their nomenclature, topology, possible structures, conserved residues, and phylogeny [13]. Accordingly, a number of tools are available for these analyses (Table 10.4). These software programs allow appropriate characterization of metagenomic sequences of the gene for a physiological function (Fig. 10.1). The majority of functional metagenomics studies characterize identified genetic loci for these features using a common set of tools [13,30].

### 10.2.2.2 In silico analysis of sequenced metagenome datasets

Sequence-based metagenomics ends with metagenomic sequence dataset of targeted genetic loci (16S rRNA gene or SSU rRNA gene) or random metagenomic datasets (Fig. 10.2). Experiments to generate these datasets were designed as per the project objective. Targeted metagenome datasets are produced to define the diversity of targeted genetic loci in the given environment. Among targeted metagenome datasets, SSU rRNA gene-based sequence datasets are common to define microbial diversity or microbial diversity associated objectives (microbial community dynamics or to identify microbial markers associated with physiological/ecological parameters) [2,7,9]. While random or whole metagenome datasets were produced to seek a holistic overview of a microbiome to answer three basic questions for any metagenomic study such as Who is in there? What are they doing? and How are they doing? [3,4,12] On the basis of the asked questions, these datasets were analyzed with different software tools.

#### 10.2.2.2.1 Quality filtration

Despite application and objective, all sequencing datasets have to be assessed for the quality parameter. Irrespective of the sequencing platform, data output is nonuniform in terms of quality score and read length. Low-quality data and short sequence reads could lead to misleading conclusions, so quality filtration is must. Accordingly, a number of software programs are available. The following are very commonly used for quality filtration of metagenomic datasets.

##### 10.2.2.2.1.1 FastQC Among various sequence quality analysis tools, FastQC is the undisputed champion of the quality control visualization tool. The FastQC tool was developed by the Babraham Institute and is available at https://www.bioinformatics.babraham.ac.uk/projects/fastqc/. It works in both stand-alone interactive systems or as a noninteractive system depending upon input file size. While FastQC does follow most of the design principles, it is still relatively labor intensive, especially when examining the quality results for large datasets [31].

**Table 10.1 Various enzymes encoding novel genes identified with the metagenomic approach.**

| Enzyme | Environmental sample |
| --- | --- |
| Cellulases (EC 3.2.1.4) | Sugarcane soil metagenome [49] |
| | Buffalo rumen metagenome [50] |
| | Soil metagenome [51] |
| Novel halotolerant cellulase | Soil metagenome [52] |
| Lipases (E.C.3.1.1.3) | Marine sediment [53] |
| | Biofilm isolated from slaughter house drain [54] |
| | Marine sponge *Ircinia* sp. [55] |
| | *Bacillus* species in an oil-contaminated habitat [56] |
| | Pond water metagenome [57] |
| | Tidal flat sediments [58] |
| | Human oral microbiome [59] |
| | Soil metagenome [60] |
| | Soil metagenome [61] |
| Novel lipolytic enzyme, EstT1 | Soil metagenome [62] |
| Salt resistant carboxylesterase | Marine metagenome [63] |
| Esterases | Soil, water metagenome [64] |
| | *Acidicaldus* sp. strain from acidic hot springs of Colombian Andes [65] |
| | China Holstein cow rumen [66] |
| Novel carboxylesterase | Soil metagenome [67] |
| Xylanase | Cow rumen metagenome [68] |
| | Compost-soil metagenome [69] |
| | Holstein cattle rumen [70] |
| | Insect gut metagenome [71] |
| Proteases | Surface sand of deserts [72] |
| | Biofilm isolated from slaughter house drain [73] |
| | Goat skin surface metagenome [74] |
| | Gobi and Death Valley deserts [75] |
| | Forest soil [76] |
| Pectinases | Soil metagenome [77] |
| | Soil from Western Ghats, India [78] |
| Novel NADP-dependent short-chain reductase | Soil metagenome, Germany [79] |
| Amylases (EC 3.2.1) | Soil metagenome [80] |
| | Soil metagenome [81] |
| Periplasmic α-amylase | Cow dung, India [82] |
| | Sponge *Arenosclera brasiliensis* microbiome [83] |
| Cold-active ß-galactosidase | Ikaite columns SW Greenland [84] |
| Halotolerant and thermostable tannase | Cotton field soil [85] |
| Salt-tolerant chitobiosidase | Agricultural soil [86] |

**Table 10.2 Novel bioactive molecules encoding genes/gene clusters identified with the metagenomics approach.**

| Compound | Environmental sample | Functional role |
|---|---|---|
| Uncharacterized protein | Soil sample from a deciduous forest, Belgium | Antimicrobial activity [87] |
| Novel lactonases | Soil metagenome, Germany | Acts as an antiinfective [88] |
| Novel chitinase | Soil metagenome, Sweden | Chitobiosidase activity [89] |
| Clone expressing NAHL-lactonase activity | Pasture soil, France | Antimicrobial activity |
| Two novel clones (SA343 and SA354) | biofilm isolated from slaughterhouse drain | Acts as Biosurfactant [90] |
| Beta-lactamase | Soil metagenome | Antibacterial activity [91] |
| | Human saliva and feces | Antibacterial activity [92] |
| Turbomycin A & B | Soil metagenome | Antibacterial activity [93] |
| Didemnin B (Aplidine) and Thiocoraline | Marine metagenome | Involved in clinical or preclinical Cancer studies [94] |
| Indirubin | Soil metagenome | Antibacterial activity [95] |
| Amikacin | Soil metagenome | Antimicrobial [96] |
| Borregomycin A | Soil metagenome from Anza-Borrego Desert in California | Has anticancer and bactericidal activities [97] |
| Patellamide and pederin | Soil metagenome | Acts as antibiotic [98] |
| Fasamycin A and B | Soil metagenome | Acts as antibiotic [99] |

**Table 10.3 Novel gene/operons involved in microbial physiology decoded with the metagenomic approach.**

| Pathways | Environment |
|---|---|
| Hypothetical protein with NF-kB pathway stimulatory activity | Human gut microbiota of Crohn's disease patients [100] |
| Novel prebiotic catabolic pathways | Human ileum mucosa and fecal microbiota samples [101] |
| Five novel putative salt tolerance genes | Human gut microbiota [102] |
| Novel salt tolerance gene | Fecal sample [103] |
| Salt resistance | Brines and moderate-salinity rhizosphere [104] |
| Salt stress tolerance | Pond water metagenome [105] |
| Arsenic resistance | Effluent treatment plant [106] |
| Acid resistance | Tinto River [107] |
| Chloramphenicol resistance | Activated sludge [108] |
| | Alluvial soil [109] |
| | Urban soil, USA [110] |
| | Agricultural soil, China [111] |
| Kanamycin resistance | Soil metagenome [112] |
| | Human fecal sample [113] |
| | MBC cheese, Italy [114] |
| Ampicillin resistance genes | Fecal samples [115] |
| | MBC cheese, Italy [114] |
| Tetracyclin resistance genes | Soil samples [116] |
| | Agricultural soil, China [117] |

**Table 10.4 Bioinformatics tools available for the characterization of the sequences identified with functional metagenomics approach.**

| Sr No. | Feature | Listed tools | Web links |
|---|---|---|---|
| 1 | Sequence assembly | DNASTAR | https://www.dnastar.com/ |
| 2 | Coding sequence or open reading frames | ORF Finder | http://www.ncbi.nlm.nih.gov/gorf |
| 3 | Homology search | BLAST | http://www.ncbi.nlm.nih.gov/Blast.cgi |
| 4 | Conserved domains | CD Search | https://www.ncbi.nlm.nih.gov/Structure/cdd/wrpsb.cgi |
| 5 | Topology | HMMTOP | http://www.enzim.hu/hmmtop/ |
| 6 | Functional annotation | PFAM | https://pfam.xfam.org/ |
| | | STRING | https://string-db.org/ |
| 7 | Sequence alignment | MUSCLE, ClustalW, T-Coffee | https://www.ebi.ac.uk/Tools/msa/muscle/ |
| | | | https://www.genome.jp/tools-bin/clustalw |
| | | | http://tcoffee.crg.cat/ |
| 8 | Signal peptide predication | Signal P | http://www.cbs.dtu.dk/services/SignalP/ |
| | | TOPCONS | http://topcons.cbr.su.se/ |
| 9 | Promoter predication | Fruitfly promoter search | http://www.fruitfly.org/seqtools/promoter.html |
| | | BPROM | http://www.softberry.com |
| 10 | Ribosomal binding sites analysis | RBS Calculator | https://salislab.net/software/ |
| 11 | Operon and transcription unit predication | fgenesB | http://www.softberry.com |
| 12. | Restriction map | Restriction Mapper | http://www.restrictionmapper.org/ |
| 13 | Phylogeny | MEGA | https://www.megasoftware.net/ |
| 14 | Nucleotide binding site | Nsite Pred Server | http://biomine.cs.vcu.edu |
| 15 | Ligand binding | RaptorX | http://raptorx.uchicago.edu/ |
| 16 | Secondary structure | Jpred4 | http://www.compbio.dundee.ac.uk/jpred/index_up.html |
| 17 | Tertiary structure | I-TASSER | https://zhanglab.ccmb.med.umich.edu/I-TASSER/ |

***10.2.2.2.1.2 Fastqp*** Fastqp is another python-based tool for quality assessment of FASTQ files, and is freely available at https://pypi.org/project/fastqp/#files. Despite high-quality graphics, it is not widely accepted [32].

***10.2.2.2.1.3 MetaQC chain*** MetaQC chain is a wide-ranging and fast quality control method for metagenomic datasets. It can be accessed at https://omictools.com/meta-qc-chain-tool. This software works in a step-wise manner starting from the technical check on input metagenomic dataset, followed by quality trimming of poor quality bases, de novo screening, and removal of contaminant sequences (generally eukaryotic sequences) from datasets. It is a very fast, accurate, freely available quality filtering tool for removal of poor quality and contaminant sequences [33].

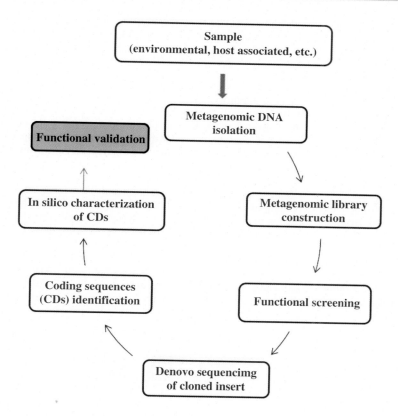

Figure 10.1 Experimental strategy of executing a functional metagenomics study.

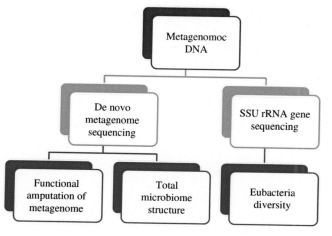

Figure 10.2 Experimental layout of a sequencing driven metagenomic study.

### 10.2.2.3    In silico analysis of 16S rRNA gene (SSU rRNA gene) datasets

Metagenomics is considered a robust method to decipher microbiome structure, function, and physiological role. On the basis of 16S rRNA gene amplification, total microbial diversity could be captured from any environment. As a standard approach, amplification of 16S rRNA gene with conserved PCR primer having unique barcode followed by sequencing of pooled samples with the high throughput sequencers was used to decipher microbial community structure. SSU rRNA gene or 16S rRNA gene is a suitable molecular marker for microbial diversity analysis of any environment. Optimized methods for direct amplification of 16S rRNA gene from metagenomic DNA followed by direct sequencing with next-generation sequencing chemistry and availability of well-established analysis sequence analysis pipelines made it the most commonly followed method to define microbiome structure (Fig. 10.3). 16S rRNA gene sequences were analyzed with different bioinformatics tools like Mothur, QIIME (Quantitative Insights Into Microbial Ecology; www.qiime.org) [11], MEGAN (http://www-ab.informatik.uni-tuebingen.de/software/megan6/), etc.

#### 10.2.2.3.1    QIIME

Quantitative Insights Into Microbial Ecology (QIIME) is a pipeline application that uses many third-party applications for analysis, interpretation, and visualization of SSU rRNA gene datasets. It is an open resource and could be employed for analysis of sequencing output of various platforms. As a part of sequencing analysis workflow, initially, it performs quality filtration, demultiplexing of SSU rRNA dataset based on the information given in map file. Chimera Slayer (http://microbiomeutil.sourceforge.net/) is employed for removal of chimera sequences. The quality-filtered sequences are clustered based on the sequence identity using any of the three options: de novo clustering (independent clustering solely based on sequence identity among themselves); open reference clustering (clustering based on homology with database sequences as well among themselves) or closed reference clustering (solely based on their identity with reference

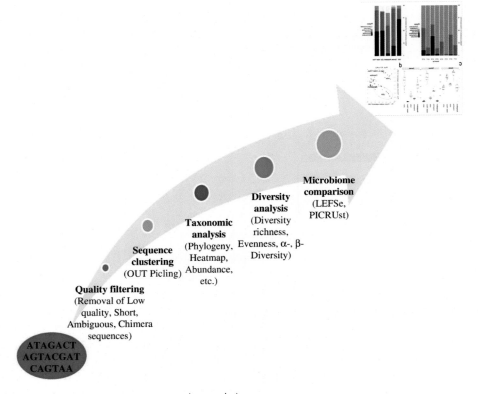

**Figure 10.3** Workflow for SSU rRNA gene sequence data analysis.

database sequences, mismatched sequences are removed from downstream processing). Each cluster has >97% sequence similarity and is referred as an OTU (operational taxonomic unit). Identified OTUs of studied samples are checked for database homology and accordingly classified in various microbial groups to calculate the distribution of SSU rRNA gene sequences among various microbial phylum, class, order, or genus. This information is utilized to develop a heat map for describing the rank abundance of microbial groups across samples. This information can be further employed for identifying microbial diversity within samples to describe microbial richness, diversity etc. Multiple samples analyzed through the QIIME pipeline can be processed with principal coordinate analysis (PCA), enterotypes identification, and core microbiome analysis. QIIME has been applied to a large number of the studies and has supported billions of sequences. QIIME allows users to initiate their analysis from raw sequencing output by OTU selection, compartmentalization assignment, construction of phylogenetic trees and production of publication-quality graphics. QIIME is, therefore, a well-accepted platform for combining heterogeneous experimental datasets and for getting new insights regarding varied microbial communities. As a result, QIIME scales to many sequences and might be used on platforms from laptops to high-performance computing clusters. QIIME is very commonly used to decipher microbiome structure, microbial community dynamics, comparative analysis of microbiome, identification of keystone microbial communities, etc. [11,13,14].

### 10.2.2.3.2 MEGAN

MEGAN (MEta Genome Analyzer) is basically the first standalone metagenome analysis software. MEGAN has been designed for studying the taxonomic content of a single environment sequence dataset. This is commonly used to define phylogenetic affiliation of sequences based on the BLAST homologs. Recently it has been updated with various additional features like COG/EGG/NOG categories analysis, sequence mapping in SEED and KEGG databases, analysis of taxonomy and function by using PCoA and publication quality visualization tools [34].

### 10.2.2.3.3 MOTHUR

Mothur (https://www.mothur.org/) is an open supply package for bioinformatics processing. Mothur is capable of processing information generated from various sequencing platforms like with 454-pyrosequencing, Illumina HiSeq and MiSeq, Sanger, PacBio, and IonTorrent. Mothur may be a bioinformatics toolkit for the requirements of the microorganism ecology connected information analysis [35].

### 10.2.2.3.4 JAGUC

JAGUC (https://omictools.com/jaguc-tool) is a Java enforced tool for microbial diversity and community structure analysis that helps to bridge the gap between computational and biological sciences. Its fast processing allows researchers to process large metagenome datasets to elucidate meaningful outcome from raw sequence data. JAGUC is a freely available standalone software package with user-defined quality filtrations and reference mapping in customized databases. It performs pairwise sequence alignment to generate clusters for rank abundance analyses [36].

### 10.2.2.3.5 UniFrac

It is an online tool (https://github.com/biocore/unifrac) for comparative microbial diversity analysis. This tool basically calculates dissimilarity score among the microbial community to define weighted and unweighted relatedness among community members. Here dissimilarity score is based distance of the members on a phylogenetic tree. It is a very quick method for comparison of microbial communities [37].

### 10.2.2.3.6 PICRUSt

Phylogenetic Investigation of Communities by Reconstruction of Unobserved States (PICRUSt) is a bioinformatics tool (http://picrust.github.io/picrust/) to predict the possible function of a microbial community based on the physiology of database homologs. It enables to decode possible functions of a microbiome without complete metagenome analysis [38].

### 10.2.2.3.7 Galaxy (https://huttenhower.sph. harvard.edu/galaxy/)

Galaxy is an open source software system using the Python programming language. Galaxy may be a scientific advancement that aims to provide a helping environment for the noncomputational biologist to perform in-depth sequence analysis without any additional programming expertise.

## 10.2.2.4 In silico analysis of whole metagenome datasets

Metagenomic fields are expanding rapidly and the data generated by metagenomic experiments are enormous and contain fragmented data representing the whole microbiome possibly having many thousand microbial species. Metagenome data is a huge dataset harboring a lot of information about a microbiome structure and its possible functions towards its survivability and functioning in context of the studied ecosystem. Accordingly curating and extracting useful biological information from these datasets is a serious bioinformatics challenge and thus a number of tools were employed for seeking meaningful biological information from metagenome datasets. There are various bioinformatics tools present for the metagenomic data analysis like sequence prefiltering, which includes the removal of redundant, low-quality sequences. Metagenome sequence assembly uses various assembly tools like Phrap, Celera, and Velvet assembler. Gene prediction metagenomic analysis is either based on homology with genes that are present in publicly sequenced databases (MEGAN4 [34]) or ab initio approach to predict coding regions based on intrinsic features of genes (GeneMark and GLIMMER).

Environmental species are very diverse and to connect community composition and function in metagenomics, sequences must be binned. Binning is the process of associating a particular sequence with an organism. This type of approach is implemented in the programs such as MEGAN [34], Phymm BL [39], MetaPhlAn [40], and in the comparative diversity analysis and species-richness utilizing tools such as Unifrac [37]. Data integration is the most important step in metagenomic data analysis as it allows the comparative analysis of different datasets using a number of ecological relationships, and for this purpose, the Metagenomic Rapid Annotation using Subsystem Technology server (MG-RAST), which is a community resource for metagenomic dataset analysis server, is present [12]. The Integrated Microbial Genomes/Metagenome (IMG/M) system also provides a collection of tools for the functional analysis of microbial communities based on their metagenomic sequence. One of the first standalone tools for analyzing high-throughput metagenomic shotgun data was MEGAN (MEtaGenome ANalyzer) [12]. This tool performs both taxonomic and functional binning, by placing the reads onto the nodes of the NCBI taxonomy using a simple lowest common ancestor (LCA) algorithm or onto the nodes of the SEED [41] or KEGG classification. All these metagenomic analysis approaches have been greatly facilitated by next-generation high throughput sequencing technologies and with the help of next-generation sequencing technology, so we are able to examine and solve various metagenomic tasks like the diversity present among the organisms that are mainly present in specific environments. Using this we can also analyze the complex interactions between members of a specific environment.

Among all, MG RAST (http://www.mcs.anl.gov/project/mg-rast-metagenomics-rast-server) is an automated pipeline for processing raw data to the metagenomic interpretation of microbiome structure and functional genetic elements, etc.

### 10.2.2.4.1 MG RAST [12]

Metagenome RAST server is an optimized pipeline for processing metagenomic data output from various next-generation sequencing platforms. Here, the user can upload a raw metagenome dataset after providing basic metadata information. After data upload, the server performs quality filtering of the uploaded dataset and removes low-quality data. Quality-filtered data are clustered based on their homology, which will eventually lead to data dereplication. This will save unnecessary data processing. Dereplicated data is searched for RNA features based on their homology with rRNA gene [42]. Identified RNA features are processed for their homology search in various databases like Greengene, RDP, etc. The rest of the metagenomic datasets are checked for the presence of coding sequences called protein features. These protein features are processed for their homology in various databases to classify these features based on COGs, NOGs, functional properties, etc. Later on, these analyses can be mapped in KEGG pathways to define possible physiological functions (Fig. 10.4). As a feature of MG RAST server, the metagenome dataset can be compared with other metagenome datasets for comparative metagenomes [12].

### 10.2.2.4.2 SEED

Straightforward Exploration of Ecological knowledge (SEED) is an efficient bioinformatics tool for exploring and visualizing microorganism community knowledge. SEED enables researchers to visually explore microbiome knowledge and it allows various analyses as the principal element and coordinate analysis (PCA/PCoA), ranked cluster, scatter plots, and bar plots. Metagenome data plots allow assessing similarities and variations among environmental samples. It is a web-based application [41].

### 10.2.2.4.3 MetaPath

It is a tool for a metabolic pathway and regulation in the metabolic pathway. MetaPath is a tool for metabolic pathway exploration and visualization [43].

## 10.3 Experimental and data analysis framework of metagenomic projects

## 10.3.1 Project I: Human gut microbiome structure

Decoding a microbiome structure is an essential and foremost step for conducting any metagenomic study. Researchers use SSU rRNA gene (or 16S rRNA gene) as a

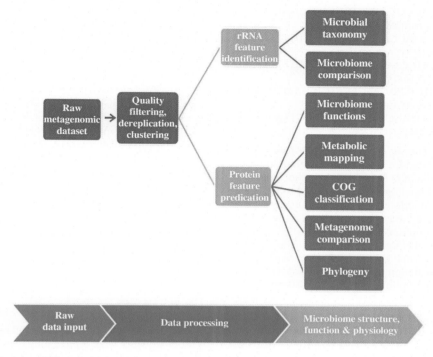

**Figure 10.4** In silico data analysis framework of a metagenomic dataset.

marker gene to decipher taxonomic distribution of microbes in the studied samples. SSU rRNA based microbial diversity analysis is commonly used for decoding microbiome structure of any environmental or host-associated sample. In this project, we will discuss a study designed to access human gut microbiome structure [44]. Experimental and analytical framework of this study is shown in a step-wise manner.

*Step 1. Isolation of Human Feces Metagenomics DNA:* Good quality, high molecular weight metagenomic DNA is the essentiality for any microbiome project. Accordingly, an optimized method was adopted to isolate metagenomic DNA.

*Step 2. Amplification and Sequencing of SSU-rRNA Gene:* SSU rRNA was amplified from human fecal metagenomic DNA using well-established 16S primers [44] following an optimized PCR condition. Amplified PCR products were directly processed for sequencing using 454 next generation sequencing methodology.

Next generation sequencing results bear lot of variability in terms of read length, quality per base, etc. Raw sequencing output appears as:

> **H7AKLNK02DBEW9 length = 47 xy = 1243_0187 region = 2 run = R_2013_04_04_13_18_49_**
ATCAGACACGGAGTTGATCCTGGCTCAGGACGAACGCT
GGCGGCGTG
> **H7AKLNK02DYHAX length = 114 xy = 1505_3207 region = 2 run = R_2013_04_04_13_18_49_**
ATCAGACACGGAGTTGATCCTGGCTCAGTACGAACGCT
GGCGGCGTGCTTAACACATGC
AAGTCGAACGAGAATCTGTGGA
ACGAGGATTCGTCCAAGGCACACAGGGAGTAG
> **H7AKLNK02DNH2F length = 78 xy = 1380_2981 region = 2 run = R_2013_04_04_13_18_49_**
ACGCTCGACAGAGTTTGATCCTGGCTCAGGATGAACGC
TGGCGGCGTGCTTAACACATGC
AAGGCACACAGGGAGTAG
> **H7AKLNK02ECOHS length = 80 xy = 1667_2158 region = 2 run = R_2013_04_04_13_18_49_**
ATATCGCGAGGAGTTTGATCCTGGCTCAGGATGAACGC
TAGCGACAGGCTTAACACATGC
AAGGCACGACAGGGGATAGG
> **H7AKLNK02C1GCE length = 301 xy = 1129_2412 region = 2 run = R_2013_04_04_13_18_49_**

AGACGCACTCGAGTTTGATCCTGGCTCAGGATGAACGC
TAGCTACAGGCTTAACACATGC
AAGTCGAGGGGAAACGATATTGGAAGCTTGCTTTTGATG
GGCGTCGACCGGCGCACGGGT
GAGTAACGCGTATCCAACCTGCCCACCACTTGGGGATA
ACCTTGCGAAAGTAAGACTAAT
ACCCAATGACGTCTCTAGAAGACATCTGAAAGAGATTAA
AGATTTATCGGTGATGGATGG
GGATGCGTCTGATTAGCTTGTTGGCGGGGTAACGGCCC
ACCAAGGCACACAGGGGAGTG

*Step 3: Quality Filtering of SSU rRNA gene:* Raw sequence
output was processed for various quality filtration
parameter (minimum Q Score > 25, Sequence length:
>250bp− < 800 bp, Primer Mismatch < 1) using "Split
Libraries" script of QIIME 1.9.0 pipeline. Quality
processed data was checked for chimeric sequences using
"Chimera Slayer" option, which finally resulted in 8802
good quality SSU rRNA gene sequences. Quality filtered
data remove short, ambiguous and chimera sequences
and add data description in sequence headers. Quality
trimming generates uniform data for downstream
processing:

> S9_1  I6Y2ZLQ03GU5RP  orig_bc = TCTCTATGCG
GA new_bc = TCTCTATGCGGA bc_diffs = 0
GACGAACGCTGGCGGCGCGCCTAACACATGCAAGT
CGAACGAGAGATGAGGAGCTTGCTCTTCAGATCGA
GTGGCGAACGGGTGAGTAACGCGTGAGGAACCTG
TCTCAAAGAGGGGNGGCAACAGTTGGAAACGACTGC
TAATACCGCATAAGCCACGGCTCGGCATCGAGCAG
AGGGAAAAGGAGTGATCCGCTTTGAGATGGCCTCG
CGTCCGATTAGCTGGTTGGTGAGGTAACGGCCCAC
CAAGC
> S7_2  I6Y2ZLQ03G96AF  orig_bc = CGTGTCTCT
AGA new_bc = CGTGTCTCTAGA bc_diffs = 0
GATGAACGCTAGCTACAGGCTTAACACATGCAAGTC
GAGGGGGCAGCATGGTCTTAGCTTGCTAAGACTGA
TGGCGACCGGCGCACGGGTGAGTAACACGTATCC
AACCTGCGTCTACTCTTGGACAGCCTTCTGAAAGGA
AGATTAATACAAGATGGCATCATGAGTCCGCATGTT
CACATGATTAAAGGTATTCCGGTAGACGATGGGGA
TGCGTTCCATTAGATAGTAGGCGGGGTAACGCCAC
CTAGTCTTCGATGGATAGGGTTCTGAGAGGAAGGT
CCCCACCCATGACATTGGAACGTAGCAGACGATCG
GTCCAAAACTTCCTTAACGGAGCAGCAGTGAGGAA
TATTGGTCAATGGGCGAGAGCCTGAACCAGCCAAAT
GCTTAGATATCACGAAGAACT
> S15_3  I6Y2ZLQ03FRUYJ  orig_bc = TCTACGTAG
CGA new_bc = TCTACGTAGCGA bc_diffs = 0
AATCAACGCTGGCGGCGTGCCTCAGACATGCAAGT
CGAACGCGAAAGTTTCCTTCGGGAGGCGAGTAGAG
TGGCGCACGGGTGAGTAACACGTAGGTCATCTACT
CTGGGTGGGAATAACAGCCGGAAACGGTTGCTAAT

ACCGCATAATACCGTAAGGTCAAAGCTTTATGTGCCT
GGAGAGGAGCCTGCGGCGGATTAGCTAGTTGGT
> S7_4  I6Y2ZLQ03HCGRU  orig_bc = CGTGTCTCT
AGA new_bc = CGTGTCTCTAGA bc_diffs = 0
GACGAACGCTGGCGGCGCGCCTAACACATGCAAGT
CGAACGAGAGATGAGGAGCTTGCTCTTCAGATCGA
GTGGCGAACGGGTGAGTAACGCGTGAGGAACCTG
CCTCAAAGAGGGAGCAACAGTTGGAAACGACTGCT
AATACCGCATAAGCCCACGGCTCGGCATCGAGCAG
AGGGAAAAGGAGTGATCCGCTTTGAGATGGCCTCG
CGTCCGATTAGCTAGTTGGTGAGGTAACGG

*Step 4: Sequence Clustering and Taxonomic Assignment of
reads:* Quality filtered data were clustered into various
groups (operational taxonomic units, OTUs) based on
their similarity ( > 97%) using UClust option. As this
project is designed to capture complete diversity, hereby
the de novo option of sequence clustering was employed
that performs ab initio OTU picking. Sequences were
assigned into an OTUs and output appears as:

```
denovo0 S5_6153
denovo1 S4_26023  S4_94601
denovo2 S15_71586 S15_102459 S15_41464 S15_487   S15_74224
denovo3 S14_97462 S14_103245 S15_3993  S14_43146
```

Here denovo 0, 1, 2, 3 are OTUs and against each OTU,
a list of sequence reads representing that OTU were listed.
After OTU identification a representative sequence was
chosen for each cluster:
> denovo0 S5_6153
GACGAACGCTGGCGGCGCGCCTAACACATGCAAG
TCGAACGAGAGAGAGGGGAGCTTGCTTCCTTGATCG
AGTGGCGAACGGGTGAGTAACGCGTGAGGAACTGC
CTCAAAGAGGGGACAACAGTTGGAAACGACTGCTA
ATACCGCATAAGCCCACGACCCGGCATCGGGAAG
AGGGAAAAGGAGCAATCCGCTTTGAGATGGCCTCG
CGTCCGATTAGCTAGTAGGTGAGGTAAAGGCTCAC
CTAGGCGACGATCTCTAGCTGGTCTGAGAGGATGA
TCAGCCACATTTGGGACTGAGACACGGCCCCAGACT
CCTACGGGAGGCAGCAGTAGGGAATATTGCACAAT
GGAGGAAACTCTGATGCAGCCATGCCGCGTGTGTG
AAGAAGGCCTTCGGGTTGTAAAGCACTTTCGGAGG
GGAGGAAAAAATTGACGTTACCCTCAGAAGAAGAC
ACCGGCTAACTCCGTGCCAGCAGCCGCGGTAATA
CGGAGGT
> denovo1 S4_26023
GACGAACGCTGGCGGCGTGCTTAACACATGCAAG
TCGAACGAAGCACTTCATAAAGTTTACTTTAAGAAG
TGACTTAGTGGCGGACGGGTGAGTAACGCGTGGG
TAACCTGCTTACACAGGGGATAACAGTTAGAAATG
ACTGCTAATACCGCATAAAACAGCAGAGTCGCATGA
CTCAACTGTCAAAGATTTATCGGTGTAAGATGGACC
CCGCGTCTGATTAGCTTGTTGGTGGGGTAACGGCC
TACCAAGGCGACGATCAGTAGCCGGCCTGAGAGGG

TGAACGGCCACATTGGGGACTGAGACACGGCCCAA
ACTCTACGGAGCAGCAGTGGGGAATATTGGACAATG
GGCGCAAGCCTGACCCAGCAACGCCGCGTGAAGG
AAGACGGTTTTCGGATTGNAAACTTCTTTTAAGAGGG
CGATAATACGGTACCTCTGATAAGCCACGGCTAACT
ACGTGCCAGCAGCCGCGG

Representative sequences from each OTU were searched for their homologs in the Greengenes database to define their taxonomic affiliation within various microbial groups. Each OTU was matched in the database and taxonomic assignments were given to each OTU at different levels and output was written in default file name "**seqs_rep_set_tax_assignments**" having complete details of matched Kingdom (k_), Phylum (p_), Class (c_), Order (o_), family (f_), Genus (g_), Species(s_) followed by their database homology.

| | | | |
|---|---|---|---|
| denovo0 | k__Bacteria; p__Bacteroidetes; c__Bacteroidia; o__Bacteroidales; f__Prevotellaceae; g__Prevotella; s__copri | 1.00 | 3 |
| denovo1 | k__Bacteria; p__Bacteroidetes; c__Bacteroidia; o__Bacteroidales; f__Bacteroidaceae; g__Bacteroides; s__fragilis | 0.67 | 3 |
| denovo2 | k__Bacteria; p__Proteobacteria; c__Gammaproteobacteria; o__Aeromonadales; f__Aeromonadaceae; g__; s__ | 0.67 | 3 |
| denovo3 | k__Bacteria; p__Proteobacteria; c__Betaproteobacteria; o__Neisseriales; f__Neisseriaceae; g__Neisseria; s__cinerea | 0.67 | 3 |

Taxonomic assignment details of each OTU were used to study microbial abundance and to construct a phylogenetic tree in .tre format that could be visualized with number of tree viewers. All this analysis has defined the microbial composition of the human gut sample.

*Step 5: Alpha Diversity analysis:* Step 4 output is in the form of a biome file, which was processed to identify species richness, evenness, and diversity within the studied sample.

*Step 6: Functional Amputation of human gut microbiome:* Biome file for step 4 was processed with Phylogenetic Investigation of Communities by Reconstruction of Unobserved States (PICRUSt) pipeline at the Galaxy server to define the physiological potential of the studied human gut sample. Functional amputation was performed for each OTU and data output is presented as:

# Constructed from biome file

| #OTU ID | S5 | S7 | S11 | S9 | S6 | S8 | S2 | S3 | S10 | S4 | S14 | S1 | S15 | S16 | S13 | S12 | KEGG_Pathways |
|---|---|---|---|---|---|---|---|---|---|---|---|---|---|---|---|---|---|
| 1,1,1-Trichloro-2,2-bis(4-chlorophenyl) ethane (DDT) degradation | 25.0 | 176.0 | 134.0 | 123.0 | 45.0 | 18.0 | 43.0 | 40.0 | 8.0 | 4.0 | 5.0 | 0.0 | 21.0 | 2.0 | 18.0 | 6.0 | Metabolism; Xenobiotics Biodegradation and Metabolism; 1,1,1-Trichloro-2,2-bis(4-chlorophenyl) ethane (DDT) degradation |
| ABC transporters | 37589.0 | 148628.0 | 47635.0 | 85585.0 | 85230.0 | 27081.0 | 98344.0 | 16036.0 | 7925.0 | 251446.0 | 72174.0 | 84651.0 | 60172.0 | 107632.0 | 69377.0 | 32442.0 | Environmental Information processing; membrane transport; ABC transporters |
| Adherens junction | 0.0 | 0.0 | 0.0 | 0.0 | 0.0 | 0.0 | 0.0 | 0.0 | 0.0 | 0.0 | 0.0 | 0.0 | 1.0 | 0.0 | 0.0 | 0.0 | Cellular ... Adherens junction |

These analyses conclude this study, which has defined its microbial diversity, rank-abundance, phylogeny, as well as potential physiology. This is a common workflow employed by a number of studies designed to decode microbial diversity or microbiome structure of soil [45], water [46], etc.

## 10.3.2 Project II Microbiome structure comparison of garden soil and hospital soil to define anthropogenic influence on soil microbiome [45]

As a part of this project, the authors have performed microbial diversity analysis of soil samples followed by microbial diversity comparison to define anthropogenic influence on soil microbiome structure. The following steps were used to attain the raised objectives:

*Step 1. Isolation of Soil Metagenomic DNA:* Good quality, high molecular weight metagenomic DNA is an essentiality for any microbiome project. Accordingly, an optimized in-house method was adopted to isolate soil metagenomic DNA.

*Step 2. Amplification and sequencing of SSU-rRNA Gene:* SSU rRNA was amplified from soil metagenomic DNA using well established bar-coded 16S primers [45] following an optimized PCR condition. Amplified PCR products were directly processed for sequencing using 454 next generation sequencing methodology.

*Step 3: Quality Filtering of SSU rRNA gene:* Raw sequence output of studied samples were processed for various quality filtration parameter (minimum Q Score > 25, Sequence length: >250 bp– <800 bp, primer mismatch <1) using "Split Libraries" script of QIIME 1.9.0 pipeline. Quality processed data was checked for chimeric sequences using "Chimera Slayer" option. Processed data output was written in default output "Seqs.fna" and data is presented as:

> HS_1 H648IQM04JK1ZT orig_bc = ATCAGACACG new_bc = ATCAGACACG bc_diffs = 0
GATGAACGCTAGCGGCAGGCCTAATACATGCAAGTCGA
ACGGTAACAGGCCCAGCAATGGGTGCTGACGAGTGGC
GCACGGGTGCGTAACACGTATGCAATCTACCTTTAACTG
GAGCATAGCCCCGAGAAATCGGGATTAATTCTCCATAGC
ATTATGAAGTGGCATCACTTTATAATTAAAGTCACAACGG
TTAAAGATGAGCATGCGCGACATTAGCTAGTTGGTGAGG
TAACGGCTCACCAAGGCTACGATGTCTAGGGGTTCTGA
GAGGATTAACCCCCACACTGGTACTGAGACACGGACCA
GACTCCTACGGGAGGCAGCAGTAAGGAATATTGGACAA
TGGTGGCAACACTGAT

> HS_2 H648IQM04JZC72 orig_bc = ATCAGACACG new_bc = ATCAGACACG bc_diffs = 0
AACGAACGCTGGCGGCGTGCCTAACACATGCAAGTCGA
ACGGAGTATTTGTAGCAATACAGATATGAGTGGCGCACG
GGTGAGTAACACGTGGATAATCTGCCTCTGAGATCGGG
ATAACCTGCCCGAAAGGCGGGCTAATACCGGATGAGGCC
ACAGTTCCGCAAGGCACACAGGGGATA

> HS_3 H648IQM04JLNTE orig_bc = ATCAGACACG new_bc = ATCAGACACG bc_diffs = 0
GACGAACGCTGGCGGTGCGCTTCATACATGCAAGTCGA
GCGATGAGGCCCTTCGGGGTACATCAGCGGCGGACGG
GTGAGTAACACGTAGATAACATGCCCTTTACTGGGGGAT
AACACCGGGAAACCGGTGCTAATACCGCATAAGTGCCA
GGCTGAAATGCCTGTCATTAAAGCTCCGGCGGTAAAGG
ATTGGTCTGCGTCTGATTAGCTAGTTGGTAGGGTAAAGG
CCTACCAAGGCTACGATCAATAGTTGGTCTGAGAGGATG
ACCAGCCACAGTGGGACTGAGACACGGCCCACACTCCT
ACGGGAGGCAGCAGTAGGGAATCTTGCGCAATGGGCG
AAAGCCTGACGCAGCGAC

> GS_1 H660VB403G5VRC orig_bc = ATATCGCGAG new_bc = ATATCGCGAG bc_diffs = 0
AGCGAACGCTGGCGGCAGGCCTAACACATGCAAGTCG
AACGCCCCGCAAGGGGAGTGGCAGACGGGTGAGTAAC
GCGTGGGAACGTACCTTTCGGTTCGGAATAACCCGGGG
AAACTCGGGCTAATACCGGATACGTCCGTGAGGAGAAA
GATTTATCGCCGAAAGATCGGCCCGCGTTGGATTAGCTA
GTTGGTGAGGTAACGGCTCACCAAGGCGATGATCCATA
GCTGGTCTGAGAGGACGATCAGCCACACTGGGACTGAG
ACACGGCCCAGACTCCTACGGGAGGCAGCAGTGGGGA
ATATTGGACAATGGGCGCAAGCCTGATCCAGCCATGCC
GCGTGAGTGATGAAGGCCTTAGGGTTGTAAAGCTCTTTT
ACCAGGGAAGATAATGACGGTACCTGGAGA

> GS_2 H660VB403FPQJ5 orig_bc = ATATCGCGAG new_bc = ATATCGCGAG bc_diffs = 0
AGCGAACGCTGGCGGCGTGCTTAACACATGCAAGTCGC
GGGAGAATGGAGGGCTTCGGCCTTCCTAGTAAACCGGC
GCACGGGTGAGTAACACGTAGGTAACCTACCTCCGAGA
CCGGGATAACCTATCGAAAGATTGGCTAATACCGGATAA
GACCACGAGGGTTTCGGCTCCTGCGGTCAAAGGTGG

> GS_3 H660VB403GO1WZ orig_bc = ATATCGCGAG new_bc = ATATCGCGAG bc_diffs = 0
GACGAACGCTGGCGGCGTGCTTAACACATGCAAGTCGA
ACGCGTGAAGCATCTTCGGGTGTGGATGAGTGGCGAAC
GGGTGAGTAACACGTGGGTAATCTGCCCTGCACTCTGG
GATAAGCCTTGGAAACGGGGTCTAATACCGGATATGACA
TTGCATCGCATGGTGTGGTGTGGAAAGTTCCGGCGGTG
CAGGATGAGCCCGCGGCCTATCAGCTTGTTGGTGGGGT
GATGGCCTACCAAGGCACACAGGGGATAGG

*Step 4: Sequence Clustering and Taxonomic Assignment of reads:* Quality filtered data were clustered into various groups (operational taxonomic units, OTUs) based on their similarity (>97%) using UClust option. As this project has been designed for comparative analysis, phylogenetic affiliation of identified OTUs is must, so closed reference OTU picking was employed. Identified OTUs were searched for their homologs in the Greengenes database to define their taxonomic affiliation within various microbial groups, which are used to study microbial abundance and to construct a phylogenetic tree.

```
# Constructed from biome file
#OTU      HS      taxonomy
ID
920243    2.0     k__Bacteria; p__Bacteroidetes;
                     c__Flavobacteria;
                     o__Flavobacteriales;
                     f__Flavobacteriaceae;
                     g__Salinimicrobium; s__
339039    2.0     k__Bacteria; p__Proteobacteria;
                     c__Alphaproteobacteria;
                     o__Rhodospirillales;
                     f__Rhodospirillaceae; g__; s__
538879    1.0     k__Bacteria; p__Proteobacteria;
                     c__Gammaproteobacteria;
                     o__Xanthomonadales;
                     f__Sinobacteraceae; g__; s__

#OTU      GS      taxonomy
ID
1124701   3.0     k__Bacteria; p__Bacteroidetes; c__
                     [Saprospirae]; o__[Saprospirales];
                     f__Chitinophagaceae; g__; s__
4479944   16.0    k__Bacteria; p__Actinobacteria;
                     c__MB-A2-108; o__0319-7L14;
                     f__; g__; s__
950120    2.0     k__Bacteria; p__Proteobacteria;
                     c__Deltaproteobacteria;
                     o__Myxococcales; f__Haliangiaceae;
                     g__; s__
```

All this analysis has defined the microbial composition of soil samples, and all these details were processed in a biome file. A Venn diagram was generated to represent common and unique OTUs in the studied samples. Even a heatmap diagram showing microbial rank distribution within the studied samples has been generated for comparative visualization (Fig. 10.5A).

*Step 5: Alpha Diversity analysis:* As we have observed different sequencing output for different samples, so the samples need to be normalized. The biome file of both samples was normalized through rarefaction followed by alpha diversity analysis of samples to define species richness, evenness, and diversity for comparative diversity analysis (Fig. 10.5B).

*Step 6: STAMP profiling:* Biome files of soil samples were processed with STAMP to identify statistically different microbial groups at the phylum, class, family, order, and

genus levels among these samples. Sample-specific microbial groups were identified from these analyses, for example, *Rhizobiales* were abundant within garden soil while the hospital soil was found enriched in *Sphingomonas* (Fig. 10.5C).

*Step 7: Functional Amputation:* Biome file for step 4 was processed with Phylogenetic Investigation of Communities by Reconstruction of Unobserved States (PICRUSt) pipeline at the Galaxy server to define the physiological potential of both soil samples. These analyses were processed to identify statistically different metabolic processes among soil samples (Fig. 10.5D).

*Step 8: Phylogenetic Tree:* A phylogenetic tree was constructed to define abundance of phylogenetically affiliated OTUs in both samples (Fig. 10.5E).

These analyses conclude this study where a comparative microbial diversity analysis of both soil samples was performed. As an outcome of the study, microbial diversity of hospital soil was less diverse, more even, having enrichment of microbial groups enriched with xenobiotic catabolic machinery as compared with the diversity of garden soil, which is highly diverse, representing nitrogen fixing, phosphate solubilizing microbes. The current study has shown a direct anthropogenic influence on soil microbiome structure. This is a common workflow employed by a number of studies designed to comparative microbial diversity analysis to identify microbial markers [47], to study microbial dysbiosis [48], etc.

### 10.3.3 Project III Metagenomic analysis of soil microbiome [45]

As a part of this project, the authors have performed metagenome analysis of soil sample to define genetic framework of the soil microbiome. The following steps were used to attain the raised objective:

*Step 1. Isolation of Soil Metagenomic DNA:* Good quality, high molecular weight metagenomic DNA was isolated with an optimized in-house method.

*Step 2. Sequencing of Metagenomic DNA:* Soil metagenomic DNA was directly processed for sequencing using 454 next generation sequencing methodology using shotgun sequencing approach(Fig. 10.6A). The sequence dataset was uploaded at MG RAST server using author login details after providing metadata information of the sample.

*Step 3: Quality Filtering of Metagenomic dataset:* Raw sequence outputs of the studied samples were checked for statistics with *DRISEE* (drisee -v -t <format> -f <input>) or *Jellyfish* (jellyfish count -C -m <6|15> -c 12 -s 1G <input>) software (Fig. 10.6B). They were further

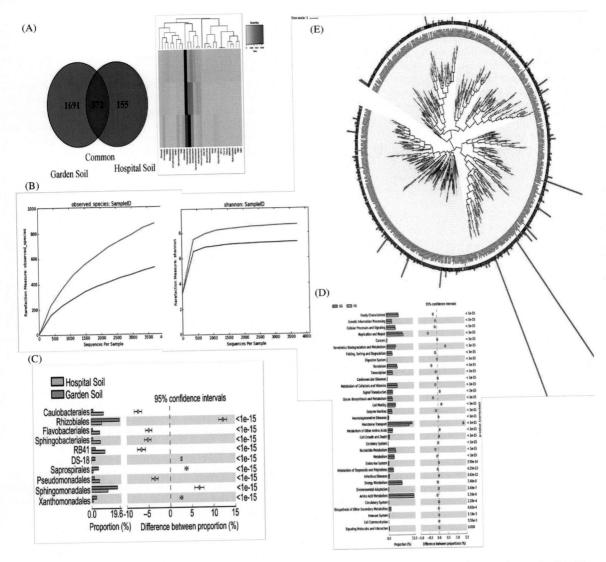

**Figure 10.5** Microbiome structure comparison of garden soil and hospital soil for distribution and rank abundance of OTUs (A), diversity estimates (B), differential microbial species (C), and differential functions (d).

processed for various quality filtration parameters like adapter trimming (autoskewer/*autoskewer.py*), denoising and normalization with *DynamicTrim* (DynamicTrimmer. pl <infile> -h <min_qual> -n <max_lqb>, and host DNA contamination removal with *Bowtie* (bowtie -f --suppress 5,6 --un <output> -t <index> <input>). The quality-filtered dataset was dereplicated to remove

redundancy for downstream analysis(pipeline/mgcmd/ *mgrast_dereplicate.pl*) (Fig. 10.6C).

*Step 4: Sequence Clustering and identification of features:* Quality filtered data were clustered based on the sequence similarity. Each cluster was identified for the presence of rRNA features using uclust option (qiime-uclust --input <input> --lib <m5rna_reduced> --uc <output> --id

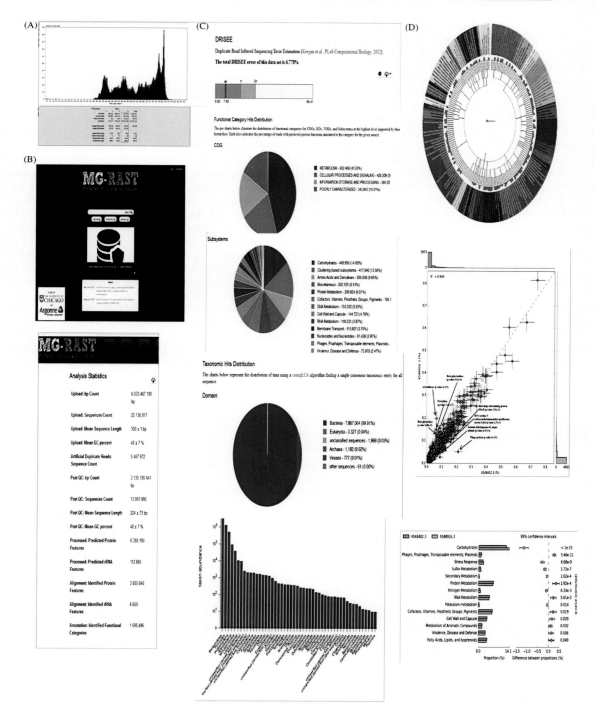

**Figure 10.6** Metagenomic analysis of soil microbiome from data output (A), analysis (B), diversity analysis (D), and metabolic pathway mapping (D).

0.75 --rev) with a cut-off of 70% identity to ribosomal sequences. These features are clustered with uclust option (iime-uclust --input <input> --uc <output> --id 0.97 --rev) with a cutoff similarity of >97%. These features were searched for their homologs in microbial diversity databases (RDP, Greengenes, ITS, etc.) with Blat software (blat -out = blast8 -t = dna -q = dna -fastMap <m5rna> <input>). Similarly putative Identify putative protein coding features (genecalling) were identified with *FragGeneScan based on* a hidden Markov model for coding regions and noncoding regions (run_FragGeneScan.pl -complete 0 -train <type> -genome <input>) followed by Filter putative protein features overlapping rRNA features (pipeline/mgcmd/*mgrast_filter_feature.pl)*. Identified protein features were clustered with uclust option (qiime-uclust --input <input> --uc <output> --id 0.90 --rev) with cutoff similarity of >90%. These were employed for Protein similarity search with BLat (blat -prot -fastMap -out = blast8 <m5nr> <input>, Protein similarity annotation (pipeline/mgcmd/*mgrast_annotate_sims.pl)*. All features were processed for Merge and index similarities (pipeline/mgcmd/*mgrast_index_sim_seq.pl)*, Annotate, and index similarities (pipeline/stages/*pipeline_loadDB)* to generate Feature abundance profile (pipeline/mgcmd/*mgrast_annotate_summary.pl)*, LCA abundance profile, Data source abundance profile (pipeline/mgcmd/*mgrast_annotate_summary.pl)*, Abundance profile load (pipeline/mgcmd/*mgrast_load_cass.pl)* (Fig. 10.6C).

*Step 5: Metabolic Mapping:* Identified protein features were mapped in the KEGG database to define physiology of the studied microbiome (Fig. 10.6D).

These analyses conclude this study performed to decipher a microbial community genome to define its structure, function, and possible physiological role [45].

## 10.4 Conclusion

Microbes harbor a complex gene pool, enabling their ubiquitous existence on Earth. Metagenomics acts as a quick doorway to explore this microbial world. Complexity and volume of metagenomic datasets is a key challenge to computational researchers, hindering them from getting a meaningful outcome from the metagenomic dataset of the studied environment. Development of various tools and various customized databases has aimed to narrow the gap between biological scientists and computational requirements. A well-defined project objective and selection of a correct data analysis pipeline is key toward the success of a metagenome project. The current chapter describes various metagenomics experimental data interpretation strategies for successful implementation of a metagenomics project.

## References

[1] Staley JT, Konopka A. Measurements of in situ activities of nonphotosynthetic microorganisms in aquatic and terrestrial habitats. Annu Rev Microbiol 1985;39:321−46.

[2] Amann RI, Ludwig W, Schleifer KH. Phylogenetic identification and in situ detection of individual microbial cells without cultivation. Microbiol Rev 1995;59 (1):143−69.

[3] Schloss PD, Handelsman J. Metagenomics for studying unculturable microorganisms: cutting the Gordian knot. Genome Biol 2005;6(8):229.

[4] Handelsman J, Rondon MR, Brady SF, Clardy J, Goodman RM. Molecular biological access to the chemistry of unknown soil microbes: a new frontier for natural products. Chem Biol 1998;5(10):245−9.

[5] Hugenholtz P, Goebel BM, Pace NR. Impact of culture-independent studies on the emerging phylogenetic view of bacterial diversity. J Bacteriol 1998;180(18):4765−74.

[6] Overmann J, Abt B, Sikorski J. Present and future of culturing bacteria. Annu Rev Microbiol 2017;71:711−30.

[7] Schloss PD, Handelsman J. Status of the microbial census. Microbiol Mol Biol Rev 2004;68 (4):686−91.

[8] Handelsman J. Sorting out metagenomes. Nat Biotechnol 2005;23 (1):38−9.

[9] Kolbert CP, Persing DH. Ribosomal DNA sequencing as a tool for identification of bacterial pathogens. Curr Opin Microbiol 1999;2(3):299−305.

[10] Schuster SC. Next-generation sequencing transforms today's biology. Nat Methods 2008;5 (1):16−18.

[11] Kuczynski J, Stombaugh J, Walters WA, González A, Caporaso JG, Knight R. Using QIIME to analyze 16S rRNA gene sequences from microbial communities. Curr Protoc Bioinformatics 2011; [chapter10: Unit 10.7].

[12] Keegan KP, Glass EM, Meyer F. MG-RAST, a metagenomics service for analysis of microbial community structure and function. Methods Mol Biol 2016;1399:207−33.

[13] Verma MK, Ahmed V, Gupta S, Kumar J, Pandey R, Mandhan V, et al. Functional metagenomics

identifies novel genes ABCTPP, TMSRP1 and TLSRP1 among human gut enterotypes. Sci Rep 2018;8(1):1397.

[14] Amann RI, Ludwig W, Schleifer KH. Phylogenetic identification and in situ detection of individual microbial cells without cultivation. Microbiol Rev 1995;59 (1):143—69.

[15] Schloss PD, Handelsman J. Biotechnological prospects from metagenomics. Curr Opin Biotechnol 2003;14(3):303—10.

[16] Pace NR, Stahl DA, Lane DJ, Olsen GJ. Analyzing natural microbial populations by rRNA sequences. ASM News 1985;51 (1):04—12.

[17] Schmidt TM, DeLong EF, Pace NR. Analysis of a marine picoplankton community by 16S rRNA gene cloning and sequencing. J Bacteriol 1991;173(14):4371—8.

[18] Riesenfeld CS, Schloss PD, Handelsman J. Metagenomics: genomic analysis of microbial communities. Annu Rev Genet 2004;38:525—52.

[19] Steele HL, Streit WR. Metagenomics: advances in ecology and biotechnology. FEMS Microbiol Lett 2005;247(2): 105—11.

[20] Handelsman J. Sorting out metagenomes. Nat Biotechnol 2005;23 (1):38—9.

[21] Thomas T, Gilbert J, Meyer F. Metagenomics — a guide from sampling to data analysis. Microb Inform Exp 2012;2(1):3.

[22] Kallifidas D, Kang HS, Brady SF. Tetarimycin A, an MRSA-active antibiotic identified through induced expression of environmental DNA gene clusters. J Am Chem Soc 2012;134 (48):19552—5.

[23] Lämmle K, Zipper H, Breuer M, Hauer B, Buta C, Brunner H, et al. Identification of novel enzymes with different hydrolytic activities by metagenome expression cloning. J Biotechnol 2007;127(4): 575—92.

[24] Heath C, Hu XP, Cary SC, Cowan D. Identification of a novel alkaliphilic esterase active at low temperatures by screening a metagenomic library from antarctic desert soil. Appl Environ Microbiol 2009;75(13):4657—9.

[25] Kapardar RK, Ranjan R, Grover A, Puri M, Sharma R. Identification and characterization of genes conferring salt tolerance to Escherichia coli from pond water metagenome. Bioresour Technol 2010;101 (11):3917—24.

[26] Chauhan NS, Ranjan R, Purohit HJ, Kalia VC, Sharma R. Identification of genes conferring arsenic resistance to Escherichia coli from an effluent treatment plant sludge metagenomic library. FEMS Microbiol Ecol 2009;67 (1):130—9.

[27] Tamaki H, Wright CL, Li X, Lin Q, Hwang C, Wang S, et al. Analysis of 16S rRNA amplicon sequencing options on the Roche/454 next-generation titanium sequencing platform. PLoS One 2011;6(9):e25263.

[28] Handelsman J. Metagenomics: application of genomics to uncultured microorganisms. Microbiol Mol Biol Rev 2004;68(4): 669—85.

[29] Lorenz P, Schleper C. Metagenome—a challenging source of enzyme discovery. J Mol Catal 2002;19-20:13—19.

[30] Ahmed V, Verma MK, Gupta S, Mandhan V, Chauhan NS. Metagenomic profiling of soil microbes to mine salt stress tolerance genes. Front Microbiol 2018;9:159.

[31] Albert. Visualizing sequencing data quality. The biostar handbook: a Beginner's guide to bioinformatics. 1st ed. Biostar; 2016.

[32] Shirley M. Simple FASTQ quality assessment using Python. Accessed v0.3. 1; 2014.

[33] Schmieder R, Edwards R. Quality control and preprocessing of metagenomic datasets. Bioinformatics 2011;27(6):863—4.

[34] Huson DH. MEtaGenomeANalyzer (MEGAN): metagenomic expert resource. Encyclopedia of metagenomics. US: Springer; 2015.

[35] Schloss PD, Westcott SL, Ryabin T, Hall JR, Hartmann M, Hollister EB, et al. Introducing mothur: open-source, platform-independent, community-supported software for describing and comparing microbial communities. Appl Environ Microbiol 2009;75 (23):7537—41.

[36] Nebel ME, Wild S, Holzhauser M, Hüttenberger L, Reitzig R, Sperber M, et al. JAGUC—a software package for environmental diversity analyses. Bioinform Comput Biol 2011;9(6):749—73.

[37] Lozupone C, Lladser ME, Knights D, Stombaugh J, Knight R. UniFrac: an effective distance metric for microbial community comparison. ISME J 2011;5 (2):169—72.

[38] Langille MG, Zaneveld J, Caporaso JG, McDonald D, Knights D, Reyes JA, et al. Predictive functional profiling of microbial communities using 16S rRNA marker gene sequences. Nat Biotechnol 2013;31(9):814—21.

[39] Brady A, Salzberg SL. Phymm and PhymmBL: metagenomic phylogenetic classification with interpolated Markov models. Nat Methods 2009;6(9):673—6.

[40] Segata N, Waldron L, Ballarini A, Narasimhan V, Jousson O, Huttenhower C. Metagenomic microbial community profiling using unique clade-specific marker genes. Nat Methods 2012;9(8): 811—14.

[41] Overbeek R, Olson R, Pusch GD, Olsen GJ, Davis JJ, Disz T, et al. The SEED and the rapid annotation of microbial genomes using subsystems technology (RAST). Nucleic Acids Res 2014; 42:206—14.

[42] Yuan C, Lei J, Cole J, Sun Y. Reconstructing 16S rRNA genes in metagenomic data. Bioinformatics 2015;31(12):35—43.

[43] Kolanczyk RC, Schmieder P, Jones WJ, Mekenyan OG, Chapkanov A, Temelkov S, et al. MetaPath: an electronic knowledge base for collating, exchanging and analyzing case studies of xenobiotic metabolism. Regul Toxicol Pharmacol 2012;63 (1):84—96.

[44] Kumar J, Kumar M, Gupta S, Ahmed V, Bhambi M, Pandey R, et al. An improved methodology to overcome key issues associated with the methods of human fecal metagenomic DNA extraction. Genomics Proteomics Bioinformatics 2016;14(6):371–8.

[45] Gupta S, Kumar M, Kumar J, Ahmad V, Pandey R, Chauhan NS. Systemic analysis of soil microbiome deciphers anthropogenic influence on soil ecology and ecosystem functioning. Int J Environ Sci Technol 2017;14(10):2229–38.

[46] Antony CP, Kumaresan D, Hunger S, Drake HL, Murrell JC, Shouche YS. Microbiology of Lonar Lake and other soda lakes. ISME J 2013;7(3):468–76.

[47] Verma MK, Ahmed V, Gupta S, Kumar J, Pandey R, Mandhan V, et al. Functional metagenomics identifies novel genes ABCTPP, TMSRP1 and TLSRP1 among human gut enterotypes. Sci Rep 2018;8(1):1397.

[48] Asghar HN, Setia R, Marschner P. Community composition and activity of microbes from saline soils and non-saline soils respond similarly to changes in salinity. Soil Biol Biochem 2012;47:175–8.

[49] Alvarez TM, Goldbeck R, dos Santos CR, Paixão DA, Gonçalves TA, Franco Cairo JP, et al. Development and biotechnological application of a novel endoxylanase family GH10 identified from sugarcane soil metagenome. PLoS One 2013;8(7):e70014.

[50] Donato JJ, Moe LA, Converse BJ, Smart KD, Berklein FC, McManus PS, et al. Metagenomic analysis of apple orchard soil reveals antibiotic resistance genes encoding predicted bifunctional proteins. Appl Environ Microbiol 2010;76(13):4396–401.

[51] Voget S, Steele HL, Streit WR. Characterization of a metagenome-derived halotolerant cellulase. J Biotechnol 2006;126(1):26–36.

[52] Garg G, Singh A, Kaur A, Singh R, Kaur J, Mahajan R. Microbial pectinases: an ecofriendly tool of nature for industries. 3 Biotech 2016;6(1):47.

[53] Peng Q, Zhang X, Shang M, Wang X, Wang G, Li B, et al. A novel esterase gene cloned from a metagenomic library from neritic sediments of the South China Sea. Microb Cell Fact 2011;10:95.

[54] Thies S, Santiago-Schübel B, Kovačić F, Rosenau F, Hausmann R, Jaeger KE. Heterologous production of the lipopeptide biosurfactant serrawettin W1 in Escherichia coli. J Biotechnol 2014;181:27–30.

[55] Slattery M, Lesser MP. Trophic ecology of sponges from shallow to mesophotic depths (3 to 150 m): comment on Pawlik et al. (2015). Mar Ecol Prog Ser 2015;527:275–9.

[56] Lee LP, Karbul HM, Citartan M, Gopinath SC, Lakshmipriya T, Tang TH. Lipase-secreting Bacillus species in an oil-contaminated habitat: promising strains to alleviate oil pollution. BioMed Res Int 2015;2015.

[57] Ranjan R, Grover A, Kapardar RK, Sharma R. Isolation of novel lipolytic genes from uncultured bacteria of pond water. Biochem Biophys Res Commun 2005;335(1):57–65.

[58] Lee MH, Lee CH, Oh TK, Song JK, Yoon JH. Isolation and characterization of a novel lipase from a metagenomic library of tidal flat sediments: evidence for a new family of bacterial lipases. Appl Environ Microbiol 2006;72(11):7406–9.

[59] Preeti A, Hemalatha D, Rajendhran J, Mullany P, Gunasekaran P. Cloning, expression and characterization of a lipase encoding gene from human oral metagenome. Indian J Microbiol 2014;54(3):284–92.

[60] Bell PJ, Sunna A, Gibbs MD, Curach NC, Nevalainen H, Bergquist PL. Prospecting for novel lipase genes using PCR. Microbiology 2002;148(8):2283–91.

[61] Voget S, Leggewie C, Uesbeck A, Raasch C, Jaeger KE, Streit WR. Prospecting for novel biocatalysts in a soil metagenome. Appl Environ Microbiol 2003;69(10):6235–42.

[62] Hosokawa M, Hoshino Y, Nishikawa Y, Hirose T, Yoon DH, Mori T, et al. Droplet-based microfluidics for high-throughput screening of a metagenomic library for isolation of microbial enzymes. Biosens Bioelectron 2015;67:379–85.

[63] Tchigvintsev A, Tran H, Popovic A, Kovacic F, Brown G, Flick R, et al. The environment shapes microbial enzymes: five cold-active and salt-resistant carboxylesterases from marine metagenomes. Appl Microbiol Biotechnol 2015;99(5):2165–78.

[64] Ouyang Z, Zheng G, Song J, Borek DM, Otwinowski Z, Brautigam CA, et al. Structure of the human cohesin inhibitor Wapl. Proc Natl Acad Sci USA 2013;110(28):11355–60.

[65] López-López O, Cerdán ME, González Siso MI. New extremophilic lipases and esterases from metagenomics. Curr Protein Pept Sci 2014;15(5):445–55.

[66] Chen W, Cai F, Zhang B, Barekati Z, Zhong XY. The level of circulating miRNA-10b and miRNA-373 in detecting lymph node metastasis of breast cancer: potential biomarkers. Tumour Biol 2013;34(1):455–62.

[67] Jeon JH, Kim SJ, Lee HS, Cha SS, Lee JH, Yoon SH, et al. Novel metagenome-derived carboxylesterase that hydrolyzes β-lactam antibiotics. Appl Environ Microbiol 2011;77(21):7830–6.

[68] Gong JS, Lu ZM, Li H, Zhou ZM, Shi JS, Xu ZH. Metagenomic technology and genome mining: emerging areas for exploring novel nitrilases. Appl Microbiol Biotechnol 2013;97(15):6603–11.

[69] Verma D, Kawarabayasi Y, Miyazaki K, Satyanarayana T. Cloning, expression and characteristics of a novel alkalistable and thermostable xylanase encoding gene (Mxyl) retrieved from compost-soil metagenome. PLoS One 2013;8(1):e52459.

[70] Cheng Y, Hung AY, Decety J. Decety Dissociation between affective

sharing and emotion understanding in juvenile psychopaths. Dev Psychopathol 2012;24(2):623–36.

[71] Brennan Y, Callen WN, Christoffersen L, Dupree P, Goubet F, Healey S, et al. Unusual microbial xylanases from insect guts. Appl Environ Microbiol 2004;70(6):3609–17.

[72] Glogauer A, Martini VP, Faoro H, Couto GH, Müller-Santos M, Monteiro RA, et al. Identification and characterization of a new true lipase isolated through metagenomic approach. Microb Cell Fact 2011;10:54.

[73] Thies S, Rausch SC, Kovacic F, Schmidt-Thaler A, Wilhelm S, Rosenau F, et al. Metagenomic discovery of novel enzymes and biosurfactants in a slaughterhouse biofilm microbial community. Sci Rep 2016;6:27035.

[74] Pushpam PL, Rajesh T, Gunasekaran P. Identification and characterization of alkaline serine protease from goat skin surface metagenome. AMB Express 2011;1 (1):3.

[75] Neveu J, Regeard C, DuBow MS. Isolation and characterization of two serine proteases from metagenomic libraries of the Gobi and Death Valley deserts. Appl Microbiol Biotechnol 2011;91 (3):635–44.

[76] Biver S, Portetelle D, Vandenbol M. Characterization of a new oxidant-stable serine protease isolated by functional metagenomics. Springerplus 2013;2:410.

[77] Singh R, Dhawan S, Singh K, Kaur J. Cloning, expression and characterization of a metagenome derived thermoactive/thermostable pectinase. Mol Biol Rep 2012;39(8):8353–61.

[78] Sathya TA, Jacob AM, Khan M. Cloning and molecular modelling of pectin degrading glycosyl hydrolase of family 28 from soil metagenomic library. Mol Biol Rep 2014;41(4):2645–56.

[79] Bijtenhoorn P, Schipper C, Hornung C, Quitschau M, Grond S, Weiland N, et al. BpiB05, a novel metagenome-derived hydrolase acting on N-acylhomoserine lactones. J Biotechnol 2011;155(1):86–94.

[80] Vidya J, Swaroop S, Singh S, Alex D, Sukumaran R, Pandey A. Isolation and characterization of a novel α-amylase from a metagenomic library of Western Ghats of Kerala, India. Biologia 2011;66 (6):939–44.

[81] Richardson TH, Tan X, Frey G, Callen W, Cabell M, Lam D, et al. A novel, high performance enzyme for starch liquefaction — discovery and optimization of a low pH, thermostable alpha-amylase. J Biol Chem 2002;277 (29):26501–7.

[82] Sharma P, Muthuirulan P, Sathyanarayanan J, Paramasamy G, Jeyaprakash R. Identification of periplasmic alpha-amylase from cow dung metagenome by product induced gene expression profiling (Pigex). Indian J Microbiol 2015;55:57–65.

[83] Silva M, Vieira L, Almeida AP, Silva A, Seca A, Barreto MC, et al. Chemical study and biological activity evaluation of two azorean macroalgae: *Ulva rigida* and *Gelidium microdon*. Oceanography 2013;1(1):102.

[84] Vester JK, Glaring MA, Stougaard P. Discovery of novel enzymes with industrial potential from a cold and alkaline environment by a combination of functional metagenomics and culturing. Microb Cell Fact 2014;13:72.

[85] Yao J, Fan XJ, Lu Y, Liu YH. Isolation and characterization of a novel tannase from a metagenomic library. J Agric Food Chem 2011;59(8):3812–18.

[86] Cretoiu MS, Berini F, Kielak AM, Marinelli F, van Elsas JD. A novel salt-tolerant chitobiosidase discovered by genetic screening of a metagenomic library derived from chitin-amended agricultural soil. Appl Microbiol Biotechnol 2015;99(19):8199–215.

[87] Biver S, Steels S, Portetelle D, Vandenbol M. Bacillus subtilis as a tool for screening soil metagenomic libraries for antimicrobial activities. J Microbiol Biotechnol 2013;23(6):850–5.

[88] Schipper C, Hornung C, Bijtenhoorn P, Quitschau M, Grond S, Streit WR. Metagenome-derived clones encoding two novel lactonase family proteins involved in biofilm inhibition in Pseudomonas aeruginosa. Appl Environ Microbiol 2009;75 (1):224–33.

[89] Hjort K, Presti I, Elväng A, Marinelli F, Sjöling S. Bacterial chitinase with phytopathogen control capacity from suppressive soil revealed by functional metagenomics. Appl Microbiol Biotechnol 2014;98(6):2819–28.

[90] Riaz K, Elmerich C, Moreira D, Raffoux A, Dessaux Y, Faure D. A metagenomic analysis of soil bacteria extends the diversity of quorum-quenching lactonases. Environ Microbiol 2008;10 (3):560–70.

[91] Allen HK, Donato J, Wang HH, Cloud-Hansen KA, Davies J, Handelsman J. Call of the wild: antibiotic resistance genes in natural environments. Nat Rev Microbiol 2010;8(4):251–9.

[92] Sommer MOA, Dantas G, Church GM. Functional characterization of the antibiotic resistance reservoir in the human microflora. Science 2009;325(5944):1128–31.

[93] Gillespie DE, Brady SF, Bettermann AD, Cianciotto NP, Liles MR, Rondon MR, et al. Isolation of antibiotics turbomycin A and B from a metagenomic library of soil microbial DNA. Appl Environ Microbiol 2002;68 (9):4301–6.

[94] Xiong ZQ, Wang JF, Hao YY, Wang Y. Recent advances in the discovery and development of marine microbial natural products. Mar Drugs 2013;11 (3):700–17.

[95] Lim HK, Chung EJ, Kim JC, Choi GJ, Jang KS, Chung YR, et al. Characterization of a forest soil metagenome clone that confers indirubin and indigo production in *Escherichia coli*. Appl Environ Microbiol 2005;71(12):7768–77.

[96] Perron GG, Whyte L, Turnbaugh PJ, Goordial J, Hanage WP, Dantas G, et al. Functional characterization of bacteria isolated from ancient arctic soil exposes diverse resistance mechanisms to modern antibiotics. PLoS One 2015;10(3):e0069533.

[97] de Castro AP, Fernandes GR, Franco OL. Insights into novel antimicrobial compounds and antibiotic resistance genes from soil metagenomes. Front Microbiol 2014;5:489.

[98] Craig JW, Chang FY, Brady SF. Natural products from environmental DNA hosted in *Ralstonia metallidurans*. ACS Chem Biol 2009;4(1):23—8.

[99] Feng Z, Chakraborty D, Dewell SB, Reddy BV, Brady SF. Environmental DNA-encoded antibiotics fasamycins A and B inhibit FabF in type II fatty acid biosynthesis. J Am Chem Soc 2012;134(6):2981—7.

[100] Lakhdari O, Cultrone A, Tap J, Gloux K, Bernard F, Ehrlich SD, et al. Functional metagenomics: a high throughput screening method to decipher microbiota-driven NF-κB modulation in the human gut. PLoS One 2010;5(9):e13092.

[101] Cecchini DA, Laville E, Laguerre S, Robe P, Leclerc M, Doré J, et al. Functional metagenomics reveals novel pathways of prebiotic breakdown by human gut bacteria. PLoS One 2013;8(9):e72766.

[102] Culligan EP, Marchesi JR, Hill C, Sleator RD. Combined metagenomic and phenomic approaches identify a novel salt tolerance gene from the human gut microbiome. Front Microbiol 2014;5:189.

[103] Culligan EP, Sleator RD, Marchesi JR, Hill C. Functional environmental screening of a metagenomic library identifies stlA; a unique salt tolerance locus from the human gut microbiome. PLoS One 2013;8(12):e82985.

[104] Mirete S, Mora-Ruiz MR, Lamprecht-Grandío M, de Figueras CG, Rosselló-Móra R, González-Pastor JE. Salt resistance genes revealed by functional metagenomics from brines and moderate-salinity rhizosphere within a hypersaline environment. Front Microbiol 2015;6:1121.

[105] Kapardar RK, Ranjan R, Grover A, Puri M, Sharma R. Identification and characterization of genes conferring salt tolerance to *Escherichia coli* from pond water metagenome. Bioresour Technol 2010;101(11):3917—24.

[106] Chauhan NS, Ranjan R, Purohit HJ, Kalia VC, Sharma R. Identification of genes conferring arsenic resistance to *Escherichia coli* from an effluent treatment plant sludge metagenomic library. FEMS Microbiol Ecol 2009;67(1):130—9.

[107] Guazzaroni ME, Morgante V, Mirete S, González-Pastor JE. Novel acid resistance genes from the metagenome of the Tinto River, an extremely acidic environment. Environ Microbiol 2013;15(4):1088—102.

[108] Parsley LC, Consuegra EJ, Thomas SJ, Bhavsar J, Land AM, Bhuiyan NN, et al. Identification of diverse antimicrobial resistance determinants carried on bacterial, plasmid, or viral metagenomes from an activated sludge microbial assemblage. Appl Environ Microbiol 2010;76(8):2673—7.

[109] Tao F, Zhang Z, Zhang S, Zhu Z, Shi W. Response of crop yields to climate trends since 1980 in China. Clim Res 2012;54:233—47.

[110] McGarvey KM, Queitsch K, Fields S. Wide variation in antibiotic resistance proteins identified by functional metagenomic screening of a soil DNA library. Appl Environ Microbiol 2012;78(6):1708—14.

[111] Su JQ, Wei B, Xu CY, Qiao M, Zhu YG. Functional metagenomic characterization of antibiotic resistance genes in agricultural soils from China. Environ Int 2014;65:09—15.

[112] Donato JJ, Moe LA, Converse BJ, Smart KD, Berklein FC, McManus PS, et al. Metagenomic analysis of apple orchard soil reveals antibiotic resistance genes encoding predicted bifunctional proteins. Appl Environ Microbiol 2010;76(13):4396—401.

[113] Cheng G, Hu Y, Yin Y, Yang X, Xiang C, Wang B, et al. Functional screening of antibiotic resistance genes from human gut microbiota reveals a novel gene fusion. FEMS Microbiol Lett 2012;336(1):11—16.

[114] Devirgiliis C, Caravelli A, Coppola D, Barile S, Perozzi G. Antibiotic resistance and microbial composition along the manufacturing process of Mozzarella di Bufala Campana. Int J Food Microbiol 2008;128(2):378—84.

[115] Martiny AC, Martiny JB, Weihe C, Field A, Ellis JC. Functional metagenomics reveals previously unrecognized diversity of antibiotic resistance genes in gulls. Front Microbiol 2011;2:238.

[116] Forsberg KJ, Reyes A, Wang B, Selleck EM, Sommer MO, Dantas G. The shared antibiotic resistance of soil bacteria and human pathogens. Science 2012;337(6098):1107—11.

[117] Su JQ, Wei B, Xu CY, Qiao M, Zhu YG. Functional metagenomic characterization of antibiotic resistance genes in agricultural soils from China. Environ Int 2014;65:09—15.

# Index

*Note*: Page numbers followed by "*f*" and "*t*" refer to figures and tables, respectively.

# Index

# Index

# Index

Printed in the United States
By Bookmasters